Do the Math
workbook

ELEMENTARY ALGEBRA
SECOND EDITION

Michael Sullivan, III
Joliet Junior College

Katherine R. Struve
Columbus State Community College

Janet Mazzarella
Southwestern College

Prentice Hall
is an imprint of

D0074617

3 4 5 6 EBM 14 13 12

ISBN-13:	978-0-321-59312-2	Standalone
ISBN-10:	0-321-59312-X	Standalone
ISBN-13:	978-0-321-59313-9	Component
ISBN-10:	0-321-59313-8	Component

Prentice Hall
is an imprint of

www.pearsonhighered.com

Do the Math
workbook

Elementary Algebra, Second Edition

Table of Contents

Five-Minute Warm-Up 1.2
Fractions, Decimals, and Percents

Definitions and vocabulary. In your own words, write a brief definition for each of the following.

1. Natural numbers _____

2. Natural numbers that are *prime* _____

3. Natural numbers that are *composite* _____

4. Least common multiple _____

5. In the statement $4 \bullet 9 = 36$,

    (a) list the factor(s).   **(b)** list the product(s).   5a. _____

                                                 5b. _____

6. List three common multiples of 8 and 12.   6. _____

7. List three common factors of 8 and 12.   7. _____

8. List all possible factors of 18.   8. _____

9. Write the prime factorization of 18.   9. _____

Guided Practice 1.2
Fractions, Decimals, and Percents

Objective 1: Factor a Number as a Product of Prime Factors

1. We use a *factor tree* to find the prime factorization of a number. The process begins with finding two factors of the given number. Continue to factor until all factors are prime.

Find the prime factorization of each natural number. (*See textbook Example 1*)

   (a) 30   **(b)** 120   1a. _____

                                                                    1b. _____

Objective 2: Find the Least Common Multiple of Two or More Numbers

2. Find the least common multiple of the numbers 8 and 12: (*See textbook Examples 2 and 3*)

Step 1: Write each number as the product of prime factors, aligning common factors vertically.

    Factor 8 as the product of primes: **(a)** _____

    Factor 12 as the product of primes: **(b)** _____

Step 2: Write down the factor(s) that the numbers share, if any. Then write down the remaining factors the greatest number of times that the factors appear in any number.

    List the common factors: **(c)** _____

    List the remaining factors: **(d)** _____

Step 3: Multiply the factors listed in Step 2. The product is the least common multiple (LCM).

    Multiply.   LCM = **(e)** _____

Objective 3: Write Equivalent Fractions

3. *Equivalent fractions* are fractions that represent the same part of a whole. Sometimes we simplify fractions, using smaller numbers to represent an equivalent part of a whole. Other times we multiply by one, written as a quantity over itself, to find larger numbers to represent the equivalent fraction.

Write the fraction $\frac{5}{8}$ as an equivalent fraction with a denominator of 24. (*See textbook Example 4*)

   (a) Determine the missing numerator and denominator to multiply by 1: $\frac{5}{8} \cdot \frac{\boxed{?}}{\boxed{?}} = \frac{}{24}$ 3a. _____

   (b) Find the equivalent fraction by determining the product: $\frac{5}{8} \cdot 1 = \frac{5}{8} \cdot \frac{\square}{\square} = \frac{?}{24}$ 3b. _____

4. Write $\dfrac{7}{36}$ and $\dfrac{11}{24}$ as equivalent fractions with the least common denominator. *(See textbook Example 5)*

Step 1: Find the least common denominator of the fractions.

Factor 36 as the product of primes: **(a)** _____

Factor 24 as the product of primes: **(b)** _____

Write the factors in the LCD: **(c)** _____

Multiply to determine the LCD: **(d)** _____

Step 2: Rewrite each fraction with the least common denominator.

To write $\dfrac{7}{36}$ on the LCD, multiply by?

(e) _____

To write $\dfrac{11}{24}$ on the LCD, multiply by?

(f) _____

Multiply to find the equivalent fractions.

(g) $\dfrac{7}{36} \cdot \dfrac{\square}{\square} =$ _____ ; $\dfrac{11}{24} \cdot \dfrac{\square}{\square} =$ _____

Objective 5: *Round Decimals*

5. It is important to identify the place value of digits in a number and to be able to round to a specific place.

Use the number 94,204.6375 to round to the required place. *(See textbook Examples 7 and 8)*

    (a) thousands   **(b)** tens   5a. _____

                                                    5b. _____

    (c) whole number (or ones)   **(d)** tenth   5c. _____

                                                      5d. _____

    (e) hundreds   **(f)** hundredth   5e. _____

                                                        5f. _____

Objective 7: *Convert a Percent to a Decimal and a Decimal to a Percent*

6. Write 75% as a decimal. *(See textbook Example 12)*   6. _____

7. Write 0.005 in percent form. *(See textbook Example 13)*   7. _____

Do the Math Exercises 1.2
Fractions, Decimals, and Percents

In Problems 1–2, find the prime factorization of each number.

1. 54

2. 63

1. _____

2. _____

In Problems 3–4, find the LCM of each set of numbers.

3. 8 and 70

4. 9, 15, and 20

3. _____

4. _____

In Problems 5–6, write each fraction with the given denominator.

5. Write $\dfrac{4}{5}$ with denominator 15.

6. Write $\dfrac{5}{14}$ with denominator 28.

5. _____

6. _____

In Problems 7–8, write the equivalent fractions with the least common denominator.

7. $\dfrac{1}{12}$ and $\dfrac{5}{18}$

8. $\dfrac{5}{12}$ and $\dfrac{7}{15}$

7. _____

8. _____

In Problems 9–10, write each fraction in lowest terms.

9. $\dfrac{9}{15}$

10. $\dfrac{24}{27}$

9. _____

10. _____

In Problems 11–12, tell the place value of the digit in the given number.

11. 9124.786; the 7

12. 539.016; the 9

11. _____

12. _____

In Problems 13–14, round each number to the given place.

13. 7298.0845 to the nearest hundred

14. 37.439 to the nearest tenth

13. _____

14. _____

In Problems 15–16, write each fraction as a terminating or repeating decimal.

15. $\dfrac{2}{9}$

16. $\dfrac{11}{32}$

15. _____

16. _____

In Problems 17–18, write each decimal as a fraction in lowest terms.

17. 0.4

18. 0.358

17. _____

18. _____

19. Write 59% as a decimal.

20. Write 0.349 as a percent.

19. _____

20. _____

Five-Minute Warm-Up 1.3
The Number Systems and the Real Number Line

In Problems 1 – 4, simplify each of the following expressions.

1. $\dfrac{0}{7}$ 2. $\dfrac{12}{0}$ 3. $\dfrac{9}{9}$ 4. $\dfrac{3}{1}$

1. _____

2. _____

3. _____

4. _____

5. In the fraction $\dfrac{3}{7}$, identify **(a)** the numerator and **(b)** the denominator.

5a. _____

5b. _____

In Problems 6 – 9, use the set of numbers $\left\{\dfrac{15}{0}, -100, \dfrac{3}{6}, 0.001, \dfrac{0}{1}, -1, \dfrac{4}{2}\right\}$. *List all the elements that are…*

6. Nonnegative   7. Positive

6. _____

7. _____

8. Positive and less than one   9. Neither positive nor negative

8. _____

9. _____

In Problems 10 and 11, write the fraction as either a repeating decimal or a terminating decimal.

10. $\dfrac{5}{6}$ 11. $\dfrac{4}{5}$

10. _____

11. _____

Guided Practice 1.3
The Number Systems and the Real Number Line

1. A *set* is a collection of objects. We use braces { } to indicate the set and enclose the objects, or *elements*, in the set. When the set has no elements in it, we say that the set is an *empty set* and use the notation ∅ or { } to indicate an empty set. It is incorrect to use {∅} to denote an empty set.

If set A is the set of digits $\{0, 1, 2, 3, 4, 5, 6, 7, 8, 9\}$, *(See textbook Example 1)*

    (a) write a set B which contains elements from set A which are even.   1a. _____

    (b) write a set C which contains elements from set A which are odd.   1b. _____

Objective 1: Classify Numbers

2. Review the definitions of the sets of numbers listed in textbook Section 1.3. These sets are:

$\mathbb{N}$ = Natural numbers W = Whole numbers $\mathbb{Z}$ = Integers
$\mathbb{Q}$ = Rational numbers Ir = Irrational numbers $\mathbb{R}$ = Real numbers

List the numbers in the set $\left\{ \dfrac{-8}{2}, 2.\overline{6}, \dfrac{0}{3}, 14, -10, \dfrac{7}{0}, -1.050050005... \right\}$ that are... *(See textbook Example 2)*

    (a) Natural _____   **(b)** Whole _____

    (c) Integer _____   **(d)** Rational _____

    (e) Irrational _____   **(f)** Real _____

3. List the sets of numbers to which the given number belongs. *(See textbook Example 2)*

    (a) π _____   **(b)** $\dfrac{12}{-6}$ _____

    (c) $1.25252525...$ _____   **(d)** $\dfrac{-100}{-20}$ _____

Objective 2: Plot Points on the Real Number Line

4. On the real number line, label the points with coordinates: $-\dfrac{5}{2}$, 3, 1.75, -0.5 *(See textbook Example 3)*

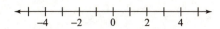

Guided Practice 1.3

Objective 3: Use Inequalities to Order Real Numbers

5. An important property of the real number line follows from the fact that given two numbers (points) a and b, either a is to the left of b ($a < b$), a is equal to b ($a = b$), or a is to the right of b ($a > b$). It is also possible that a is less than or equal to b $(a \leq b)$ or a is greater than or equal to b $(a \geq b)$.

Replace the question mark by <, >, or =, whichever is correct. *(See textbook Example 4)*

    (a) $-10 \ ? -10.5$ **(b)** $\dfrac{3}{4} ? \, 0.75$ **(c)** $\dfrac{8}{6} \, ? \, \dfrac{18}{15}$ 5a. _____

5b. _____

5c. _____

6. Determine whether the statement is True or False.

    (a) $0 > -2$ **(b)** $5 \geq \dfrac{-10}{-2}$ **(c)** $-5 \leq -5.\overline{1}$ 6a. _____

6b. _____

6c. _____

7. If x represents any real number, how would you use inequality symbols to write

    (a) x is positive?   **(b)** x is negative?   7a. _____

7b. _____

Objective 4: Compute the Absolute Value of a Real Number

8. The *absolute value* of a number a, written $|a|$, is the distance from 0 to a on the real number line. Because distance is always positive, the absolute value of a any non-zero number is always

8. _____

9. Evaluate each of the following: *(See textbook Example 5)*

    (a) $\left| \dfrac{0}{-5} \right|$ **(b)** $|-4.2|$ $|120|$ 9a. _____

9b. _____

9c. _____

10   Sullivan/Struve/Mazzarella, *Elementary Algebra*, 2e

Do the Math Exercises 1.3
The Number Systems and the Real Number Line

In Problems 1–3, write each set.

1. *B* is the set of *natural* numbers less than 25.

1. _____

2. *C* is the set of *integers* between –6 and 4, not including –6 or 4.

2. _____

3. *F* is the set of odd *natural* numbers less than 1.

3. _____

In Problems 4–6, list the elements in the set $\left\{-4, 3, -\dfrac{13}{2}, 0, 2.303003000...\right\}$ *that are described.*

4. whole numbers   **5.** rational numbers

4. _____

5. _____

6. real numbers

6. _____

7. Plot the points in the set $\left\{\dfrac{3}{4}, \dfrac{0}{2}, -\dfrac{5}{4}, -0.5, 1.5\right\}$ on a real number line.

In Problems 8–9, determine whether the statement is True or False.

8. $0 < -5$ **9.** $-3 \geq -5$

8. _____

9. _____

In Problems 10–12, replace the ? with the correct symbol: $>, <, =$.

10. $-8 \ ? \ -8.5$ **11.** $\dfrac{5}{12} \ ? \ \dfrac{2}{3}$

10. _____

11. _____

12. $\dfrac{5}{11} \ ? \ 0.\overline{45}$

12. _____

In Problems 13–15, evaluate each expression.

13. $|-8|$ **14.** $|7|$

13. _____

14. _____

15. $\left| -\dfrac{13}{9} \right|$

15. _____

Five-Minute Warm-Up 1.4
Adding, Subtracting, Multiplying, and Dividing Integers

In Problems 1 – 4, what operation is indicated by each of the following words?

1. Quotient   **2.** Difference   1. _____

                                                                                          2. _____

3. Sum   **4.** Product   3. _____

                                                                                          4. _____

5. Write an expression for 9 divided by 18 using each of the following symbols:

(a) $\dfrac{a}{b}$ **(b)** a/b **(c)** $a \div b$ **(b)** $a\overline{)b}$ 5a. _____

                                                                                          5b. _____

                                                                                          5c. _____

                                                                                          5d. _____

6. Evaluate each of the following:

(a) $\left| -13 \right|$ **(b)** $\left| \dfrac{15}{4} \right|$ **(c)** $\left| 0 \right|$ 6a. _____

                                                                                          6b. _____

                                                                                          6c. _____

7. Write $\dfrac{90}{36}$ as a fraction in lowest terms.   7. _____

Guided Practice 1.4
Adding, Subtracting, Multiplying, and Dividing Integers

Objective 1: Add Integers

1. To add two integers that are the same sign, add their absolute values.

(a) If both numbers are positive, the sign of the sum is _____.   1a. _____

(b) If both numbers are negative, the sign of the sum is _____.   1b. _____

2. To add two integers that are different signs, subtract the absolute value of the smaller number from the

absolute value of the larger number. The sign of the sum is _____.

3. Find the sum: $-45 + (-22)$ *(See textbook Example 5)*

Step 1: Add the absolute values of the two integers.

Find the absolute value of $|-45|$: (a) _____

Find the absolute value of $|-22|$: (b) _____

Add (a) and (b): (c) _____

Step 2: Attach the common sign, either positive or negative.

The sign of both numbers is negative so the sign of the sum is: (d) _____

Write the sum: (e) _____

4. Find the sum: $73 + (-81)$ *(See textbook Example 6)*

Step 1: Subtract the smaller absolute value from the larger absolute value.

Find the absolute value of $|73|$: (a) _____

Find the absolute value of $|-81|$: (b) _____

Subtract (a) from (b): (c) _____

Step 2: Attach the sign of the integer with the larger absolute value.

Which number has the larger absolute value: (d) _____

What is the sign of (d): (e) _____

Therefore, the sign of the sum is: (f) _____

Write the sum: (g) _____

Objective 2: Determine the Additive Inverse of a Number

5. For any real number a other than 0, there is a real number $-a$, called the *additive inverse*, or *opposite*, of a, having the property: $a + (-a) = -a + a = 0$.

Determine the additive inverse of the given real number. *(See textbook Example 7)*

(a) $\dfrac{9}{2}$

(b) -10.3

5a. _____

5b. _____

Objective 3: Subtract Integers

6. Find the difference: $-32 - (-61)$ *(See textbook Example 8)*

Step 1: Change the subtraction problem to an equivalent addition problem.

Write the equivalent addition problem:

(a) _____

Step 2: Find the sum.

Use the rules from Objective 1 to find the sum from (a):

(b) _____

Objective 4: Multiply Integers

7. To multiply two integers, multiply the numbers without regard to the sign.

(a) If both numbers have the same sign, the sign of the product is _____.

7a. _____

(b) If both numbers have different signs, the sign of the product is _____.

7b. _____

(c) _____ is neither positive nor negative.

7c. _____

8. Find the product. *(See textbook Example 11)*

(a) $6(-12)$

(b) $(-4)(-15)$

(c) $-13 \bullet 9$

8a. _____

8b. _____

8c. _____

Objective 5: Divide Integers

In Problems 9 – 10, identify the (a) dividend, (b) divisor, and (c) the quotient.

9. $\dfrac{36}{12}$

10. $(-72) \div 72$

9a. _____ 9b. _____ 9c. _____

10a. _____ 10b. _____ 10c. _____

11. Find the multiplicative inverse (reciprocal) of -3. *(See textbook Example 14)*

11. _____

Sullivan/Struve/Mazzarella, *Elementary Algebra*, 2e

Do the Math Exercises 1.4
Adding, Subtracting, Multiplying, and Dividing Integers

In Problems 1–4, find the sum.

1. $-4 + 12$

2. $-13 + (-5)$

1. _____

2. _____

3. $-145 + (-68)$

4. $(-13) + 37 + (-22)$

3. _____

4. _____

In Problems 5–6, determine the additive inverse of each real number.

5. -34

6. 7

5. _____

6. _____

In Problems 7–9, find the difference.

7. $12 - 19$

8. $-15 - 9$

7. _____

8. _____

9. $46 - (-25)$

9. _____

In Problems 10–13, find the product.

10. $7 \bullet 9$

11. $(-22)(-5)$

10. _____

11. _____

12. $(-128)7$

13. $-6 \bullet 4 \bullet 8$

12. _____

13. _____

In Problems 14–15, find the multiplicative inverse (or reciprocal) of each number.

14. 10

15. -3

14. _____

15. _____

In Problems 16–18, find the quotient.

16. $36 \div 9$

17. $\dfrac{-144}{6}$

16. _____

17. _____

18. $\dfrac{-80}{-12}$

18. _____

In Problems 19–20, write each expression using mathematical symbols. Then evaluate the expression.

19. the sum of 32 and –64

20. –40 divided by 100

19. _____

20. _____

Five-Minute Warm-Up 1.5

Adding, Subtracting, Multiplying, and Dividing Rational Numbers Expressed as Fractions and Decimals

In Problems 1 – 4, perform the indicated operations.

1. $(-15)(-6)$

2. $\dfrac{125}{-5}$

1. _____

2. _____

3. $-36 - 42$

4. $27 + (-163)$

3. _____

4. _____

5. Write $\dfrac{24}{96}$ as a fraction in lowest terms.

5. _____

6. Find the least common denominator of $\dfrac{4}{15}$ and $\dfrac{5}{24}$.

6. _____

7. Rewrite $\dfrac{5}{12}$ as an equivalent fraction with a denominator of 60.

7. _____

Guided Practice 1.5
Adding, Subtracting, Multiplying, and Dividing Rational Numbers Expressed as Fractions and Decimals

1. Write each rational number in lowest terms.

(a) $\dfrac{-12}{15}$ (b) $-\dfrac{24}{3}$ (c) $\dfrac{-5}{-25}$ 1a. _____

1b. _____

1c. _____

Objective 1: Multiply Rational Numbers Expressed as Fractions

2. Find the product: $-\dfrac{7}{72} \cdot \dfrac{81}{35}$ *(See textbook Example 1)*   2. _____

Objective 2: Divide Rational Numbers Expressed as Fractions

3. Find the quotient: $\dfrac{8}{15} \div \dfrac{4}{45}$ *(See textbook Example 2)*

Step 1: Write the equivalent multiplication problem.

$$\dfrac{8}{15} \div \dfrac{4}{45} =$$

(a) _____

Step 2: Write the product in factored form and divide out common factors.

Write the prime factorization of the numerator and denominator:

(b) _____

Divide out common factors:

(c) _____

Step 3: Multiply the remaining factors.

$$\dfrac{8}{15} \div \dfrac{4}{45} =$$

(d) _____

Objective 3: Add and Subtract Rational Numbers Expressed as Fractions

4. Adding or subtracting fractions with the same denominator use the following properties:

$$\dfrac{a}{c} + \dfrac{b}{c} = \dfrac{a+b}{c},\, c \neq 0 \qquad \dfrac{a}{c} - \dfrac{b}{c} = \dfrac{a-b}{c},\, c \neq 0$$

Find the difference and write in lowest terms, if necessary: $\dfrac{7}{12} - \dfrac{11}{12}$ *(See textbook Example 4)*

(a) Use the property $\dfrac{a}{c} - \dfrac{b}{c} = \dfrac{a-b}{c}$ to write the difference over a single denominator: 4a. _____

(b) Rewrite as an addition problem: 4b. _____

(c) Add the numerators: 4c. _____

(d) Factor and divide out common factors. State the difference: $\dfrac{7}{12} - \dfrac{11}{12} =$ 4d. _____

5. Find the sum: $\dfrac{5}{12} + \dfrac{4}{9}$ *(See textbook Example 5)*

Step 1: Find the least common denominator of the denominators.

Write 12 as the product of primes: **(a)** _____

Write 9 as the product of primes: **(b)** _____

Write the factors in the LCD: **(c)** _____

Multiply to determine the LCD: **(d)** _____

Step 2: Write each rational number with the denominator found in Step 1.

To write $\dfrac{5}{12}$ on the LCD, multiply by? **(e)** _____

To write $\dfrac{4}{9}$ on the LCD, multiply by? **(f)** _____

Multiply to find the equivalent fractions.

(g) $\dfrac{5}{12} \cdot \dfrac{\square}{\square} = $; $\dfrac{4}{9} \cdot \dfrac{\square}{\square} = $

Step 3: Add the numerators and write the result over the common denominator.

Add the fractions: **(h)** _____

Step 4: Write in lowest terms.

The result is already in lowest terms. State the sum $\dfrac{5}{12} + \dfrac{4}{9}$: **(i)** _____

6. Evaluate and write in lowest terms, if necessary: $-3 + \dfrac{5}{8}$ *(See textbook Example 7)*

 (a) Write the integer -3 as a fraction: 6a. _____

 (b) Write each fraction as an equivalent fraction over the common denominator: 6b. _____

 (c) Add the numerators and write the result over the common denominator: 6c. _____

 (d) Write the fractions in lowest terms, if necessary. State the sum of $-3 + \dfrac{5}{8}$:   6d. _____

Objective 4: Add, Subtract, Multiply, and Divide Rational Numbers Expressed as Decimals

7. To add or subtract decimals, we arrange the numbers in a column with the decimals aligned. Place the decimal point in the answer directly below the decimal point in the problem. When adding or subtracting

decimals, any number that has no decimal point has an implied decimal located _____.

8. The number of digits to the right of the decimal point in the product is the same as the _____ of the number of digits to the right of the decimal point in the factors.

Do the Math Exercises 1.5
Adding, Subtracting, Multiplying, and Dividing Rational Numbers Expressed as Fractions and Decimals

In Problems 1–4, find the product, and write in lowest terms.

1. $\dfrac{7}{8} \cdot \dfrac{10}{21}$

2. $-\dfrac{3}{7} \cdot 63$

3. $-\dfrac{5}{2} \cdot \dfrac{16}{25}$

4. $-\dfrac{60}{75} \cdot \left(-\dfrac{25}{36}\right)$

1. _____

2. _____

3. _____

4. _____

In Problems 5–6, find the reciprocal of each number.

5. $\dfrac{9}{4}$

6. -8

5. _____

6. _____

In Problems 7–10, find the quotient, and write in lowest terms.

7. $\dfrac{1}{2} \div \dfrac{3}{6}$

8. $-\dfrac{1}{4} \div 4$

9. $\dfrac{4}{3} \div \left(-\dfrac{9}{10}\right)$

10. $\dfrac{44}{63} \div \dfrac{11}{21}$

7. _____

8. _____

9. _____

10. _____

In Problems 11–14, find the sum or difference and write in lowest terms.

11. $\dfrac{6}{11} + \dfrac{16}{11}$ **12.** $-\dfrac{7}{8} + 4$

11. _____

12. _____

13. $-\dfrac{2}{5} + \left(-\dfrac{2}{3}\right)$ **14.** $-\dfrac{29}{6} - \left(-\dfrac{29}{20}\right)$

13. _____

14. _____

In Problems 15–20, perform the indicated operation.

15. $-(-32.9) + 10.3$ **16.** $32 - 5.68$

15. _____

16. _____

17. $3.1 \bullet 10.9$ **18.** $0.065 \bullet 340$

17. _____

18. _____

19. $\dfrac{332.59}{7.9}$ **20.** $\dfrac{48}{0.03}$

19. _____

20. _____

Five-Minute Warm-Up 1.6
Properties of Real Numbers

In Problems 1 – 4, perform the indicated operations.

1. $-5 + 5 + (-2)$ 1. _____

2. $\dfrac{5}{9} \bullet \dfrac{9}{5} \bullet (-12)$ 2. _____

3. $\dfrac{8}{1} \bullet \dfrac{1}{12}$ 3. _____

4. $-33 \bullet \left(-\dfrac{15}{11}\right)$ 4. _____

5. $\dfrac{-75}{-15}$ 5. _____

Guided Practice 1.6
Properties of Real Numbers

Objective 1: Understand and Use the Identity Properties of Addition and Multiplication

1. Based on the stated property, write the right side of the equation.

    (a) $\dfrac{4}{7} + 0 = $ —— ; Identity Property of Addition   1a. _____

    (b) $\dfrac{9}{9} \bullet \left(-\dfrac{1}{4}\right) = $ —— ; Multiplicative Identity   1b. _____

2. Perform the following conversions. *(See textbook Example 1)*

    (a) 198 inches = ? feet [1 foot = 12 inches]   2b. _____

    (b) 15 minutes = ? seconds [1 minute = 60 seconds]   2b. _____

Objective 2: Understand and Use the Commutative Properties of Addition and Multiplication

3. The commutative properties of addition and multiplication say that the _____ in which a   3. _____
problem is written will not change the result.

4. The following operations are *not* commutative: _____ and _____

5. Based on the stated property, write the right side of the equation. *(See textbook Example 2)*

    (a) $(-12) + 6 = $ _____ ; Commutative Property of Addition   5a. _____

    (b) $(4 + 20) \bullet \dfrac{1}{2} = $ _____ ; Commutative Property of Multiplication   5b. _____

6. Evaluate the following expressions. *(See textbook Examples 3 and 4)*

    (a) $(-17) + 4 + 17$ 6a. _____

    (b) $42 \bullet 5 \bullet \left(\dfrac{3}{7}\right)$ 6a. _____

Objective 3: Understand and Use the Associative Properties of Addition and Multiplication

7. The associative properties of addition and multiplication say that if a problem contains   7. _____
only addition or contains only multiplication, we can keep the order the same but change the
_____ of the numbers in the problem and the result will not change.

8. Based on the stated property, write the right side of the equation. *(See textbook Example 5)*

    (a) $(-9 + 12) + 2 =$ _____ ; Associative Property of Addition   8a. _____

    (b) $3 \bullet \left(\dfrac{2}{3} \bullet 15 \right) =$ _____ ; Associative Property of Multiplication   8b. _____

9. Use an Associative Property to evaluate the following expressions. *(See textbook Examples 6 and 7)*

    (a) $(-7) + 17 + 10$ 9a. _____

    (b) $12 \bullet \left(\dfrac{5}{6} \bullet 5 \right)$ 9a. _____

Objective 4: Understand the Multiplication and Division Properties of 0
10. Complete the following. *(See textbook Example 8)*

    (a) $a \bullet 0 = 0 \bullet a =$ _____   **(b)** $\dfrac{0}{a} =$ _____; $a \neq 0$ **(c)** $\dfrac{a}{0} =$ _____; $a \neq 0$

Do the Math Exercises 1.6
Properties of Real Numbers

In Problems 1–4, convert each measurement to the indicated unit of measurement. Use the following conversions:

1 gallon = 4 quarts   3 feet = 1 yard
16 ounces = 1 pound   100 centimeters = 1 meter

1. 130 feet to yards   **2.** 5900 centimeters to meters   1. _____

                                                                                                          2. _____

3. 58 quarts to gallons   **4.** 120 ounces to pounds   3. _____

                                                                                                          4. _____

In Problems 5–8, state the property of real numbers that is being illustrated.

5. $4 \bullet 63 \bullet \frac{1}{4} = 4 \bullet \frac{1}{4} \bullet 63$ **6.** $(4 \bullet 5) \bullet 7 = 4 \bullet (5 \bullet 7)$ 5. _____

                                                                                                          6. _____

7. $\frac{-8}{0}$ **8.** $\frac{5}{12} \bullet \frac{12}{5} = 1$ 7. _____

                                                                                                          8. _____

In Problems 9–17, evaluate each expression by using the properties of real numbers.

9. $46 + 59 + (-46)$ **10.** $\frac{4}{9} \bullet \frac{9}{4} \bullet 28$ 9. _____

                                                                                                          10. _____

11. $36 \cdot (-12) \cdot \dfrac{1}{6}$

12. $\dfrac{13}{2} \cdot \dfrac{8}{39} \cdot \dfrac{39}{4}$

11. _____

12. _____

13. $\dfrac{0}{100}$

14. $4000(0.5)(0.001)$

13. _____

14. _____

15. $104 \cdot \dfrac{1}{104}$

16. $30 \cdot \dfrac{4}{4}$

15. _____

16. _____

17. $\dfrac{7}{48} \cdot \left(-\dfrac{21}{4}\right) \cdot \dfrac{12}{7}$

17. _____

Sullivan/Struve/Mazzarella, *Elementary Algebra,* 2e

Five-Minute Warm-Up 1.7
Exponents and the Order of Operations

1. Find the sum: $-56 + (-25)$

1. _____

2. Find the difference: $-17 - 31$

2. _____

3. Find the product: $-36 \cdot \left(\dfrac{8}{9} \right)$

3. _____

4. Find the quotient: $\dfrac{135}{-3}$

4. _____

5. Evaluate the expression: $35 - 43 + (-18) + 96$

5. _____

6. Find the product: $(-3) \cdot (-3) \cdot (-3) \cdot (-3)$

6. _____

Guided Practice 1.7
Exponents and the Order of Operations

Objective 1: Evaluate Exponential Expressions

1. Integer exponents provide a shorthand device for representing repeated multiplications of a real number. In the expression 2^5,

(a) 2 is called the _____

1a. _____

(b) 5 is called the _____

1b. _____

(c) To evaluate the expression, multiply _____ = _____.

1c. _____

(See textbook Example 2)

2. Identify the base for each expression. *(See textbook Example 1)*

(a) -9^2 (b) $(-9)^4$

2a. _____

2b. _____

3. Consider the number -6.

(a) Another word for raising the number -6 to the second power is to say -6 ___.

3a. _____

(b) This is written _____.

3b. _____

(c) To evaluate the expression, multiply _____ = _____.

3c. _____

(See textbook Example 4)

4. Consider the number -5.

(a) Another word for raising the number -5 to the third power is to say -5 ____.

4a. _____

(b) This is written _____.

4b. _____

(c) To evaluate the expression, multiply _____ = _____.

4c. _____

(See textbook Example 3)

Objective 2: Apply the Rules for Order of Operations

5. Complete the following table.

		Order of Operations	
(a)	**P**	_____	This includes all possible grouping symbols.
(b)	**E**	_____	Work from left to right.
(c)	**M/D**	_____	Work from left to right.
(d)	**A/S**	_____	Work from left to right.

6. We use a variety of grouping symbols. When multiple pairs of grouping symbols exist and are nested inside one another, we evaluate the information in the innermost grouping symbol first and work our way outward.

Besides nested parentheses, what grouping symbols might appear in an expression with multiple operations?

6. _____

7. Evaluate the expression: $5 - 24 \div 6 \bullet 9$ *(See textbook Example 5)*

7. _____

8. Evaluate the expression: $\dfrac{4 \bullet 2^3 + (-16)}{3(4-6)^2}$ *(See textbook Examples 9 and 10)*

Step 1: Evaluate the expression in parentheses first.

$\dfrac{4 \bullet 2^3 + (-16)}{3(4-6)^2}$

= **(a)** _____

Step 2: Evaluate the exponents.

= **(b)** _____

Step 3: Find products.

= **(c)** _____

Step 4: Add terms in numerator.

= **(d)** _____

Step 5: Write in lowest terms.

= **(e)** _____

Do the Math Exercises 1.7
Exponents and the Order of Operations

In Problems 1–2, write in exponential form.

1. $4 \bullet 4 \bullet 4 \bullet 4 \bullet 4$ **2.** $(-8)(-8)(-8)$ 1. _____

2. _____

In Problems 3–8, evaluate each exponential expression.

3. 2^5 **4.** $\left(\dfrac{5}{2}\right)^4$ 3. _____

4. _____

5. $(0.04)^2$ **6.** -5^4 5. _____

6. _____

7. 1^6 **8.** $\left(-\dfrac{3}{2}\right)^5$ 7. _____

8. _____

In Problems 9–19, evaluate each expression.

9. $12 + 8 \bullet 3$ **10.** $50 \div 5 \bullet 4$ 9. _____

10. _____

11. $(7-5)\bullet\dfrac{5}{2}$

12. $\dfrac{5+3}{3+15}$

11. _____

12. _____

13. $12 - [7 + (-6)3]$

14. $(-11.8 - 15.2) \div (-2)$

13. _____

14. _____

15. $-10 - 4^2$

16. $-5^2 + 3^2 \div (3^2 + 9)$

15. _____

16. _____

17. $\left(\dfrac{7-5^2}{8+4\bullet 2}\right)^2$

18. $3\bullet\left[6\bullet(5-2)-2\bullet 5\right]$

17. _____

18. _____

19. $\left(\dfrac{3}{4}+\dfrac{1}{2}\right)\left(\dfrac{2}{3}-\dfrac{1}{2}\right)$

19. _____

Sullivan/Struve/Mazzarella, *Elementary Algebra*, 2e

Five-Minute Warm-Up 1.8
Simplifying Algebraic Expressions

1. Find the sum: $-14 + 37$ 1. _____

2. Find the difference: $-26 - (-16)$ 2. _____

3. Find the product: $72 \bullet \left(-\dfrac{5}{8}\right)$ 3. _____

4. Find the quotient: $\dfrac{-105}{-15}$ 4. _____

5. Evaluate each of the following:
   (a) $(-4)^2$ **(b)** -4^2 5a. _____

                                                                                            5b. _____

6. Simplify: $\left(-13 + (-2)\right)^2$ 6. _____

Guided Practice 1.8
Simplifying Algebraic Expressions

Objective 1: Evaluate Algebraic Expressions

1. A letter that represents any number from a set of numbers is called a(n) _____.

1._____

2. A fixed number, such as 3, or a letter or symbol that represents a fixed number is a(n) _____.

2._____

3. Any combination of variables, constants, grouping symbols, and mathematical operations is called a(n) _____ _____.

3._____

4. When we substitute a numerical value for each variable in an expression and then simplify the result, we are _____ an algebraic expression.

4._____

5. Evaluate the algebraic expression $2x^2 - 5x + 3$ for $x = -2$. *(See textbook Example 1)*

5._____

Objective 2: Identify Like Terms and Unlike Terms

6. A constant or the product of a constant and one or more variables raised to a power is called a _____.

6._____

7. The numerical factor of a term is called a(n) _____.

7._____

8. When terms have the same variable(s) and the same exponent(s) on the variables, we say that the expressions are _____ _____.

8._____

9. Identify the terms in the algebraic expression $x^2 - 2xy - y^2$. *(See textbook Example 3)*

9._____

10. Determine the coefficient of each term. *(See textbook Examples 4 and 5)*

    (a) $\dfrac{x}{2}$ **(b)** $-y$ **(c)** 14   **(d)** $-\dfrac{2z}{5}$

10a._____

10b._____

10c._____

10d._____

11. Determine if the following pairs of terms are *like* or *unlike*. *(See textbook Example 6)*

    (a) $4xy^2$ and $2x^2y$ **(b)** $\dfrac{3}{4}n^3$ and $5n^3$

11a. _____

11b. _____

Objective 3: Use the Distributive Property

12. Use the Distributive Property to remove the parentheses. *(See textbook Example 7)*

    (a) $2(3x-1)$ **(b)** $-\dfrac{1}{5}(10y+5)$

12a. _____

12b. _____

Objective 4: Simplify Algebraic Expressions by Combining Like Terms

13. Combine like terms. *(See textbook Examples 8 and 9)*

    (a) $x+3x$ **(b)** $4x^2-x+2-4x^2+7x+1$

13a. _____

13b. _____

14. Simplify the algebraic expression: $5+2(6x+1)-(3x-5)$ *(See textbook Example 10)* 14. _____

Do the Math Exercises 1.8
Simplifying Algebraic Expressions

In Problems 1–2, evaluate each expression using the given values of the variables.

1. $n^2 - 4n + 3$ for $n = 2$ **2.** $-2p^2 + 5p + 1$ for $p = -3$ 1. _____

                                                                         2. _____

In Problems 3–4, for each expression, identify the terms and then name the coefficient of each term.

3. $3m^4 - m^3n^2 + 4n - 1$ **4.** $t^3 - \dfrac{t}{4}$ 3. _____

                                                                         4. _____

In Problems 5–6, determine if the terms are like or unlike.

5. -13 and 38 **6.** x^2y^3 and y^2x^3 5. _____

                                                                         6. _____

In Problems 7–8, use the Distributive Property to remove the parentheses.

7. $3(4s + 2)$ **8.** $-5(k - n)$ 7. _____

                                                                         8. _____

In Problems 9–14, simplify each expression by using the Distributive Property to remove parentheses and combining like terms.

9. $x + 2y + 5x + 7y$

10. $-7p^5 + 2p^5$

9. _____

10. _____

11. $-(-6m + 9n - 8p)$

12. $18m - (6 + 9m)$

11. _____

12. _____

13. $\dfrac{3}{5}y + \dfrac{7}{10}y$

14. $\dfrac{1}{5}(60 - 15x) + \dfrac{3}{4}(12 - 24x)$

13. _____

14. _____

15. Ticket Sales A community college theatre group sold tickets to a recent production. Student tickets cost \$5 and nonstudent tickets cost \$8. The algebraic expression $5s + 8n$ represents the total revenue from selling s student tickets and n nonstudent tickets. Evaluate $5s + 8n$ for $s = 76$ and $n = 63$.

15. _____

Sullivan/Struve/Mazzarella, *Elementary Algebra*, 2e

Five-Minute Warm-Up 2.1
Linear Equations: The Addition and Multiplication Properties of Equality

1. Determine the additive inverse of $-\dfrac{2}{3}$.

    1. _____

2. Determine the multiplicative inverse of 14.

    2. _____

3. Evaluate: $-\dfrac{5}{2}\left(-\dfrac{2}{5}\right)$

    3. _____

4. Use the Distributive Property to simplify: $-3(6x+5)$

    4. _____

5. Simplify: $9-\dfrac{1}{2}(12x-4)$

    5. _____

6. Evaluate $\dfrac{7}{4}x+2$ for $x=-16$.

    6. _____

Guided Practice 2.1
Linear Equations: The Addition and Multiplication Properties of Equality

Objective 1: Determine If a Number Is a Solution of an Equation

1. How do you determine if a value of a variable is a solution to an equation?

2. Determine if the given value of the variable is a solution of the equation $-3x + 5 = -10$.
(See textbook Example 1)

   (a) $x = 5$ **(b)** $x = -5$ 2a. _____

                                                                           2b. _____

Objective 2: Use the Addition Property of Equality to Solve linear Equations

3. In your own words, state the Addition Property of Equality.

4. Will the Addition Property of Equality allow you to subtract the same number from both sides of an equation to form an equivalent equation? Explain your reasoning.

5. The goal in solving a linear equation is to get the variable by itself with a coefficient of 1. We call this

process _____ _____ _____.

6. Solve the linear equation: $x + 13 = -11$ *(See textbook Example 2)*

Step 1: Isolate the variable x on the left side of the equation.		$x + 13 = -11$
	Subtract 13 from both sides:	**(a)** $x + 13 - \underline{\hspace{0.6cm}} = -11 - \underline{\hspace{0.6cm}}$
Step 2: Simplify the left and right sides of the equation.	Apply the Additive Inverse Property, $a + (-a) = 0$:	**(b)** _____
	Apply the Additive Identity Property, $a + 0 = a$	**(c)** _____
Step 3: Check Verify your value for x in the original equation.	Replace x in the original equation to see if a true statement results.	$x + 13 = -11$
	State the solution set:	**(d)** _____

Objective 3: Use the Multiplication Property of Equality

7. In your own words, state the Multiplication Property of Equality. Do you believe this property will hold true for division as well? Why or why not?

8. Solve the linear equation: $-9x = 324$ *(See textbook Example 5)*

Step 1: Get the coefficient of the variable x to be 1.		$-9x = 324$
	Multiply both sides by the reciprocal of the coefficient, $-\dfrac{1}{9}$:	**(a)** _____ $\bullet (-9x) =$ _____ $\bullet (324)$
Step 2: Simplify the left and right sides of the equation.	Regroup factors using the Associative Property of Multiplication:	**(b)** _____
	Apply the Multiplicative Inverse Property, $a \bullet \dfrac{1}{a} = 1$:	**(c)** _____
	Apply the Multiplicative Identity, $1 \bullet a = a$:	**(d)** _____
Step 3: Check Verify your value for x in the original equation.	Replace x in the original equation to see if a true statement results.	$-9x = 324$
	State the solution set:	**(e)** _____

9. State the first step necessary to isolate the variable. *(See textbook Examples 6 – 9)*

(a) $3p = -45$

9a. _____

(b) $\dfrac{x}{15} = -15$

9b. _____

(c) $-18 = -\dfrac{2}{3}y$

9c. _____

(d) $\dfrac{8}{3} = \dfrac{4}{9}n$

9d. _____

Do the Math Exercises 2.1
Linear Equations: The Addition and Multiplication Properties of Equality

In Problems 1–4, determine if the given value is a solution to the equation. Answer Yes or No.

1. $4t + 2 = 16$; $t = 3$ **2.** $3(x + 1) - x = 5x - 9$; $x = -3$

1. _____

2. _____

3. $-15 = 3x - 16$; $x = \dfrac{1}{3}$ **4.** $3s - 6 = 6s - 3.4$; $s = -1.2$

3. _____

4. _____

In Problems 5–8, solve the equation using the Addition Property of Equality. Be sure to check your solution.

5. $13 = u - 6$ **6.** $-2 = y + 13$

5. _____

6. _____

7. $x - \dfrac{1}{8} = \dfrac{3}{8}$ **8.** $\dfrac{3}{8} = y - \dfrac{1}{6}$

7. _____

8. _____

In Problems 9–14, solve the equation using the Multiplication Property of Equality. Be sure to check your solution.

9. $4z = 30$

10. $-8p = 20$

9. _____

10. _____

11. $\dfrac{4}{3}b = 16$

12. $-\dfrac{6}{5}n = -36$

11. _____

12. _____

13. $\dfrac{1}{4}w = \dfrac{7}{2}$

14. $\dfrac{3}{10}q = -\dfrac{1}{6}$

13. _____

14. _____

15. New Kayak The total cost for a new kayak is $862.92, including sales tax of $63.92. To find the cost of the kayak without tax, solve the equation $k + 63.92 = 862.92$, where k represents the cost of the kayak.

15. _____

Five-Minute Warm-Up 2.2
Linear Equations: Using the Properties Together

1. Simplify by combining like terms: $3x - 2(5x + 4) - 4x$ 1. _____

2. Evaluate the expression $-7(4x - 3) - 12$ for $x = -3$.   2. _____

3. Simplify: $\dfrac{4}{3} \bullet \left(\dfrac{3}{4}x \right)$ 3. _____

4. Simplify by combining like terms: $8x + 3 - 12 - 6x$ 4. _____

5. Simplify: $\dfrac{-12}{-28}$ 5. _____

6. Divide: $\dfrac{106}{0.8}$ 6. _____

Guided Practice 2.2
Linear Equations: Using the Properties Together

Objective 1: Apply the Addition and the Multiplication Properties of Equality to Solve Linear Equations

1. Solve the equation: $6x - 9 = -27$ *(See textbook Example 1)*

Step 1: Isolate the term containing the variable.

Apply the Addition Property of Equality and add ___ to both sides of the equation:

$6x - 9 = -27$

(a) _____

(b) $6x - 9 +$ ___ $= -27 +$ ___

Simplify:

(c) _____

Step 2: Get the coefficient of the variable to be 1.

Apply the Multiplication Property of Equality and divide both sides of the equation by ___ (this is the same as multiplying both sides by ___):

(d) _____ ; _____

(e) $\dfrac{6x}{\underline{}} = \dfrac{-18}{\underline{}}$

Simplify:

(f) _____

Step 3: Check Verify your value for x in the original equation.

Replace x in the original equation to see if a true statement results.

$6x - 9 = -27$

State the solution set:

(g) _____

2. Solve the equation: $\dfrac{5}{4}n + 3 = -12$ *(See textbook Example 2)*

$\dfrac{5}{4}n + 3 = -12$

   (a) Subtract 3 from both sides of the equation:

(a) _____

   (b) Simplify:

(b) _____

   (c) Multiply both sides of the equation by $\dfrac{4}{5}$:

(c) _____

   (d) Simplify:

(d) _____

   (e) State the solution set:

(e) _____

Objective 2: Combine Like Terms and Apply the Distributive Property to Solve Linear Equations

3. Before using the Addition or Multiplication Property of Equality to solve an equation, we must begin by simplifying each side of the equation. If a side of the equation has two or more like terms,

we begin by _____

4. When an equation contains parentheses, we use the _____ Property to remove the parentheses before we use the Addition or Multiplication Property of Equality to solve the equation.

5. In the following equations, combine like terms and/or use the Distributive Property to remove the parentheses. What equation is left to solve? *DO NOT SOLVE THE EQUATION.*
(See textbook Examples 3 and 4)

(a) $5x + 3 - 7x + 2 = -4$ **(b)** $4 + 8(3n + 4) = -2(n - 1) - 3$ 5a. _____

5b. _____

Objective 3: Solve a Linear Equation with the Variable on Both Sides of the Equation

6. Our goal is to get the terms that contain the variable on one side of the equation and the constant terms on the other side. In the equation $6 - 5x = 4 + 3x$, what steps would accomplish this goal? **Hint:** There is more than one option. *(See textbook Example 5)*

7. Solve the equation: $3(z + 5) - 16z = 2 - (z + 3)$ *(See textbook Example 6)*

Step 1: Remove any parentheses using the Distributive Property.		$3(z + 5) - 16z = 2 - (z + 3)$
	Use the Distributive Property:	**(a)** _____
Step 2: Combine like terms on each side of the equation.	Combine like terms:	**(b)** _____
Step 3: Use the Addition Property of Equality to get the terms with the variable on one side of the equation and the constants on the other side.	Add z to both sides of the equation:	**(c)** _____
	Simplify:	**(d)** _____
	Subtract 15 from both sides:	**(e)** _____
	Simplify:	**(f)** _____
Step 4: Use the Multiplication Property of Equality to get the coefficient of the variable to be 1.	Divide both sides by -12:	**(g)** _____
	Simplify:	**(h)** _____
Step 5: Check the solution to verify that is satisfies the original equation.	We leave the check to you.	
	State the solution set:	**(i)** _____

Do the Math Exercises 2.2
Linear Equations: Using the Properties Together

In Problems 1–4, solve the equation. Check your solution.

1. $-4x + 3 = 15$ **2.** $6z - 7 = 3$ 1. _____

                                                                    2. _____

3. $\dfrac{5}{4}a + 3 = 13$ **4.** $\dfrac{1}{5}p - 3 = 2$ 3. _____

                                                                    4. _____

In Problems 5–8, solve the equation. Check your solution.

5. $5r + 2 - 3r = -14$ **6.** $2b + 5 - 8b = 23$ 5. _____

                                                                    6. _____

7. $3(t - 4) = -18$ **8.** $-5(6 + z) = -20$ 7. _____

                                                                    8. _____

In Problems 9–12, solve the equation. Check your solution.

9. $3 + 8x = 21 - x$ **10.** $6 - 12m = -3m + 3$ 9. _____

                                                                    10. _____

11. $-4(10 - 7x) = 3x + 10$

12. $-8 + 4(p + 6) = 10p$

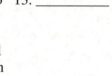

13. Wendy's A Wendy's Mandarin Chicken salad contains 10 grams of fat less than a Taco Bell Zesty Chicken Border Bowl. If there are 60 grams of fat in the two salads, find the number of grams of fat in the Taco Bell Zesty Chicken Border Bowl by solving the equation $x + (x - 10) = 60$, where x represents the number of grams of fat in a Taco Bell Border Bowl and $x - 10$ represents the number of grams of fat in the Wendy's Mandarin Chicken salad.

14. Overtime Pay Juan worked a total of 50 hours last week and earned \$498.75. He earned 1.5 times his regular hourly rate for 6 hours, and double his hourly rate for 4 holiday hours. Solve the equation $40x + 6(1.5x) + 4(2x) = \$498.75$, where x is Juan's regular hourly rate.

15. Perimeter of a Triangle The perimeter of a triangle is 210 inches. If the sides are made up of 3 consecutive even integers, find the lengths of each of the 3 sides by solving the equation $x + (x + 2) + (x + 4) = 210$, where x represents the length of the shortest side.

Sullivan/Struve/Mazzarella, *Elementary Algebra*, 2e

Five-Minute Warm-Up 2.3
Solving Linear Equations Involving Fractions and Decimals; Classifying Equations

1. Find the LCD of $\dfrac{6}{7}$ and $\dfrac{2}{5}$.

1. _____

2. Find the LCD of $-\dfrac{4}{25}$ and $\dfrac{7}{45}$.

2. _____

3. Use the Distributive Property to simplify: $-4\left(\dfrac{3}{4}x - 1\right)$

3. _____

4. Multiply: $100(0.43x - 2.7)$

4. _____

5. Simplify by combining like terms: $3 - 0.5(10x - 50) + 6x$

5. _____

6. Simplify by combining like terms: $\dfrac{4}{3}\left[9 + 3(x - 5)\right]$

6. _____

7. Simplify by combining like terms: $6\left(\dfrac{5x + 1}{3} - 2\right)$

7. _____

Guided Practice 2.3
Solving Linear Equations Involving Fractions and Decimals; Classifying Equations

Objective 1: Use the Least Common Denominator to Solve a Linear Equation Containing Fractions

1. Solve the linear equation: $\dfrac{3}{2}x - \dfrac{4}{3}x = -\dfrac{9}{4}$ *(See textbook Example 1)*

Step 1: Apply the Distributive Property to remove parentheses.	Determine the LCD:	**(a)** _____
	Use the Multiplication Property of Equality:	**(b)** ___ $\bullet \left(\dfrac{3}{2}x - \dfrac{4}{3}x\right) =$ ___ $\bullet \left(-\dfrac{9}{4}\right)$
	Use the Distributive Property:	**(c)** ___ $\bullet \left(\dfrac{3}{2}x\right) -$ ___ $\bullet \left(\dfrac{4}{3}x\right) =$ ___ $\bullet \left(-\dfrac{9}{4}\right)$
	Simplify (c):	**(d)** _____
Step 2: Combine like terms.	Combine like terms:	**(e)** _____
Step 3: Use the Addition Property of Equality to get the terms with the variable on one side of the equation and the constants on the other side.	This step is not necessary as the variable term is on the left side and the constant term on right.	
Step 4: Get the coefficient of the variable to be 1.	Divide both sides by 2:	**(f)** _____
	Simplify:	**(g)** _____
Step 5: Check Verify your value for *x* in the original equation.	Replace *x* in the original equation to see if a true statement results.	
	State the solution set:	**(h)** _____

2. Solve the equation $\dfrac{4x+3}{9} + 1 = \dfrac{2x+1}{2}$. *(See textbook Example 2)*

We will multiply both sides of the equation by the LCD to remove the fractions from the equation. *DO NOT FINISH SOLVING THE EQUATION.*

(a) Identify the LCD of the denominators in the equation: **(a)** _____

(b) Apply the Multiplication Property of Equality. **(b)** _____

(c) Distribute the LCD to each terms and divide out common factors: **(c)** _____

(d) Use the Distributive Property to remove parentheses: **(d)** _____

(e) Combine like terms on each side of the equation: **(e)** _____

Objective 2: Solve a Linear Equation Containing Decimals

3. When encountering an equation that contains decimals, an option is to multiply both sides of the equation by some power of 10 that will clear the decimals. Try solving an equation that contains decimals using this technique and alternately leaving the decimals in the equation. Which method do you prefer?

Solve: $0.06x + 0.075(7500 - x) = 517.5$ *(See textbook Example 5)*   3. _____

Objective 3: Classify a Linear Equation as an Identity, Conditional, or a Contradiction

4. Solve the linear equation $-2(3x - 1) = 4 - 6(x + 3)$. State whether the equation is an identity, contradiction or conditional equation. *(See textbook Example 6)*

    (a) Is the last statement of the solution true or false? _____

    (b) Therefore the solution set is _____

    (c) Is this equation an identity, contradiction or a conditional equation? _____

Objective 4: Use Linear Equations to Solve Problems

5. Martina saves dimes and quarters in a piggy bank. She opened the bank and discovered that she had $24.85 from a total of 127 coins. Martina wants to know how many dimes and how many quarters are in the bank….apparently, she doesn't want to just count them. To find the number of quarters, q, she solves the equation: $0.10(127 - q) + 0.25q = 24.85$. *(See textbook Example 9)*

    (a) Solve the equation. What value did you find for the value of q? _____

    (b) Check. Number of dimes plus number of quarters = 127? _____ Total value of $24.85? _____

    (c) Answer the question: How many dimes were there?_____ How many quarters were there?_____

Do the Math Exercises 2.3
Solving Linear Equations Involving Fractions and Decimals; Classifying Equations

In Problems 1–4, solve the equation. Check your solution.

1. $\dfrac{3}{2}n - \dfrac{4}{11} = \dfrac{91}{22}$

2. $\dfrac{3m}{8} - 1 = \dfrac{5}{6}$

1. _____

2. _____

3. $\dfrac{3}{2}b - \dfrac{4}{5}b = \dfrac{28}{5}$

4. $\dfrac{2}{3}(6 - x) = \dfrac{5x}{6}$

3. _____

4. _____

In Problems 5–8, solve the equation. Check your solution.

5. $0.3z = 6$

6. $p + 0.04p = 260$

5. _____

6. _____

7. $0.7y - 4.6 = 0.4y - 2.2$

8. $5 - 0.2(m - 2) = 3.6m + 1.6$

7. _____

8. _____

In Problems 9–12, solve the equation. State whether the equation is a contradiction, an identity, or a conditional equation.

9. $4(y - 2) = 5y - (y + 1)$

10. $-3x + 2 + 5x = 2(x + 1)$

9. _____

10. _____

11. $7b + 2(b - 4) = 8b - (3b + 2)$

12. $\dfrac{2m+1}{4} - \dfrac{m}{6} = \dfrac{m}{3} - 1$

11. _____

12. _____

13. Team Sweatshirt Your favorite college has logo sweatshirts on sale for $33.60. The sweatshirt has been marked down by 30%. To find the original price of the sweatshirt, x, solve the equation $x - 0.30x = 33.60$.

13. _____

14. Clean Car Pablo cleaned out his car and found nickels and quarters in the car seats. He found $4.25 in change and noticed that the number of quarters was 5 less than twice the number of nickels. Solve the equation $0.05n + 0.25(2n - 5) = 4.25$ to find n, the number of nickels Pablo found.

14. _____

15. Paying Your Taxes You are married and just determined that you paid $12,200 in federal income taxes in 2008. The solution to the equation $12,200 = 0.15(x - 14,600) + 1460$ represents the adjusted gross income of you and your spouse in 2008. Determine the adjusted gross income of you and your spouse in 2008.

15. _____

Five-Minute Warm-Up 2.4
Evaluating Formulas and Solving Formulas for a Variable

1. Evaluate the expression $2L + 2W$ for $L = 10.5$ and $W = 3.5$

1. _____

2. Round the expression 12.3762 to the hundredths place.

2. _____

3. Write the percent as a decimal: 7.25%

3. _____

4. Multiply and round your answer to the nearest tenth: $25 \bullet \pi$

4. _____

5. Solve: $-4x + 5 = -11$

5. _____

Guided Practice 2.4
Evaluating Formulas and Solving Formulas for a Variable

Objective 1: Evaluate a Formula

1. A formula is an equation that describes how two or more variables are related. If you travel in Europe and the temperature is $20°$ Celsius, use the formula $F = \dfrac{9}{5}C + 32$ to find the equivalent temperature in degrees Fahrenheit. *(See textbook Example 1)*

1. _____

2. Many times we solve problems that involve simple interest. Simple interest is money that is paid for the use of money, without any effect of compounding (receiving interest on the interest). For each of the variables in the simple interest formula, identify what the variable represents.

$I = Prt$			
I		r	
P		t	

3. List the formulas for each of the following geometric figures. *(Refer to textbook Section 2.4, as necessary.)*

(a) Square	**(b) Rectangle**
Area:_____ Perimeter: _____	Area:_____ Perimeter: _____
(c) Triangle	**(d) Trapezoid**
Area:_____ Perimeter: _____	Area:_____ Perimeter: _____
(e) Parallelogram	**(f) Circle**
Area:_____ Perimeter: _____	Area:_____ Circumference: _____
(g) Cube	**(h) Rectangular Solid**
Volume:_____ Surface Area: _____	Volume:_____ Surface Area: _____
(i) Sphere	**(j) Right Circular Cylinder**
Volume:_____ Surface Area:_____	Volume:_____ Surface Area: _____
(k) Cone	
Volume:_____	

4. Find the number of meters of fencing that must be purchased to enclose a dog run that is 8.5 meters long and 6.75 meters wide. If fencing cost $15 per meter, how much will it cost to install the fence?

   (a) What formula will you need to solve this problem? *(See textbook Examples 4 and 5)* 4a. _____

   (b) Substitute the values in the problem into the formula to find how much fencing is required.

4b. _____

   (c) Multiply to find the cost.

4c. _____

Objective 2: Solve a formula for a Variable

5. Solve $6x - 12y = -24$ for y. *(See textbook Examples 7 and 8)*

Step 1: Isolate the term containing the variable, y.	Subtract $6x$ from both sides of the equation to isolate the term $-12y$:	$6x - 12y = -24$ **(a)** _____
	Simplify:	**(b)** _____
Step 2: Get the coefficient of the variable to be 1.	Divide both sides of the equation by -12:	**(c)** _____
	Simplify using $\dfrac{a-b}{c} = \dfrac{a}{c} - \dfrac{b}{c}$:	**(d)** $y = $ _____

6. The amount of profit P, earned by a manufacturer is given by the formula $P = R - C$, where R represents the manufacturer's revenue and C represents the manufacturer's costs. *(See textbook Example 9)*

   (a) Solve the equation for C, the manufacturer's costs.   6a. _____

   (b) Find the amount of costs incurred if the manufacturer has a $5200 profit and $7300   6b. _____
   in revenue.

      Sullivan/Struve/Mazzarella, *Elementary Algebra*, 2e

Do the Math Exercises 2.4
Evaluating Formulas and Solving Formulas for a Variable

In Problems 1–5, substitute the given values into the formula and then evaluate to find the unknown quantity. Label units in the answer. If the answer is not exact, round your answer to the nearest hundredth.

1. **Length of a Bridge** The formula $m = 0.3048f$ converts length in feet f to length in meters m. The George Washington Bridge in New York City is 3500 feet long. How long is the George Washington Bridge in meters?

   1.

2. **Salesperson's Earnings** The formula $E = 750 + 0.07S$ is a formula for the earnings, E, of a salesperson who receives $750 per week plus 7% commission on all sales, S. Find the earnings of a salesperson who had weekly sales of $1200.

   2. _____

3. **Investing an Inheritance** Christopher invested his $5000 inheritance from his grandmother in a 9-month Certificate of Deposit that earns 4% simple interest per annum. Use the formula $I = Prt$ to find the amount of interest Christopher's investment will earn.

   3. _____

4. Consider the rectangle below.

   20
   32

   a) Find the perimeter of the rectangle.

   4a. _____

   (b) Find the area of the rectangle.

   4b. _____

5. For the problems below, use $\pi \approx 3.14$.

   $r = 2.8$

   (a) Find the circumference of the circle.

   5a. _____

   (b) Find the area of the circle.

   5b. _____

In Problems 6–9, solve each formula for the stated variable.

6. $F = mv^2$; solve for m

7. $V = \dfrac{1}{3} Bh$; solve for B

6. _____

7. _____

8. $S = a + b + c$; solve for b

9. $P = 2l + 2w$; solve for l

8. _____

9. _____

In Problems 10–13, solve each formula for y.

10. $-2x + y = 18$

11. $12x - 6y = 18$

10. _____

11. _____

12. $5x + 6y = 18$

13. $\dfrac{2}{3}x - \dfrac{5}{2}y = 5$

12. _____

13. _____

In Problems 14–15, (a) solve for the indicated variable, and then (b) find the value of the unknown quantity. When given, label units in the answer.

14. Profit = Revenue – Cost: $P = R - C$

(a) Solve for R.

14a. _____

(b) Find R when $P = \$4525$ and $C = \$1475$.

14b. _____

15. Simple Interest: $I = Prt$

(a) Solve for t.

15a. _____

(b) Find t when $I = \$42$, $P = \$525$, and $r = 4\%$.

15b. _____

Five-Minute Warm-Up 2.5
Introduction to Problem Solving: Direct Translation Problems

1. Solve the equation: $x - 17.38 = -72.1$ 1. _____

2. Solve the equation: $x + 0.05x = 33.6$ 2. _____

3. Each of the following words or phrases can be replaced with one of the symbols: $+ \: / \: - \: / \: \times \: / \: \div$.
   Select the correct symbol for each word or phrase.

(a) twice	**(b)** less than	**(c)** of	**(d)** sum	3a. _____
				3b. _____
				3c. _____
				3d. _____
(e) per	**(f)** increased by	**(g)** difference	**(h)** quotient	3e. _____
				3f. _____
				3g. _____
				3h. _____
(i) fewer	**(j)** greater than	**(k)** ratio	**(l)** decreased by	3i. _____
				3j. _____
				3k. _____
				3l. _____
(m) product	**(n)** exceeds by	**(o)** more than	**(p)** double	3m. _____
				3n. _____
				3o. _____
				3p. _____
(q) less	**(r)** times	**(s)** half	**(t)** altogether	3q. _____
				3r. _____
				3s. _____
				3t. _____

Guided Practice 2.5
Introduction to Problem Solving: Direct Translation Problems

Objective 1: Translate English Phrases to Algebraic Expressions

1. Express each English phrase as an algebraic expression. *(See textbook Example 1)*

    (a) The sum of -5 and $3x$ **(b)** 5 times the sum of r and 25   1a. _____

     1b. _____

    (c) The sum of 3 times a number x and 12   **(d)** The ratio of twice a number p and 7   1c. _____

     1d. _____

    (e) 18 less than half of a number z **(f)** The difference of w and 900   1e. _____

     1f. _____

2. Order matters when subtracting and dividing. Translate the English phrase as an algebraic expression.

    (a) a less than b **(b)** a less b **(c)** a subtracted from b 2a. _____

     2b. _____

     2c. _____

    (d) a divided by b **(e)** a divided into b 2d. _____

     2e. _____

Objective 2: Translate English Sentences to Equations

3. Translate each sentence into an equation. Do not solve the equation. *(See textbook Example 4)*

    (a) The sum of 5 and x is 19.   **(b)** 14 less than twice y is 30.   3a. _____

     3b. _____

    (c) The difference of z and 9 is the same as 6 more than two-fifths of z.   3c. _____

4. List six steps for solving problems with mathematical models.

_____   _____

_____   _____

_____   _____

Objective 3: Build Models for Solving Direct Translation Problems

5. If n represents the first of three unknown consecutive integers, how would you represent the next two

    integers? **(a)** _____   **(b)** _____

6. If x represents the first of four unknown consecutive even integers, how would you represent the next

    three even integers? **(a)** _____   **(b)** _____   **(c)** _____

7. If *p* represents the first of four unknown consecutive odd integers, how would you represent the next three

odd integers? **(a)** _____ **(b)** _____ **(c)** _____

8. The sum of three consecutive odd integers is 453. Find the integers. *(See textbook Example 6)*

> **Step 1: Identify** We are looking for three consecutive odd integers whose sum is 453.

> **(a) Step 2: Name** Let *n* represent the first integer. What expression represents the next

>> odd integer? _____ the third odd integer? _____

> **(b) Step 3: Translate** Write an equation for the sum of the three odd integers. _____

> **(c) Step 4: Solve** your equation from Step 3. What value did you find for *n*? _____

> **Step 5: Check** Find the other integers. Do they meet the criteria in the problem?

> **(d) Step 6: Answer the Question** _____

9. Lucky you! A long-lost uncle left you $20,000 in his will. You have decided to invest the money in two
different accounts, a conservative bond fund and a more aggressive stock account. If you will invest $5000
less in the bond fund than the stock account, how much of the $20,000 will you invest in each account?
(See textbook Example 8)

> **Step 1: Identify** We want to know how much to invest in each account.

> **(a) Step 2: Name** Let *x* represent the amount in the stock account. What algebraic expression

>> represents $5000 less going in the bond fund? _____

> **(b) Step 3: Translate** Write an equation for the sum of the two accounts: _____

> **(c) Step 4: Solve** your equation from Step 3. What value did you find for *x*? _____

> **Step 5: Check** Use your value for the stock account and Step 2 to find how much is in the bond
> account. Do these answers seem reasonable? Do they meet all of the criteria?

> **(d) Step 6: Answer the Question** _____

Do the Math Exercises 2.5
Introduction to Problem Solving: Direct Translation Problems

In Problems 1–4, translate each phrase to an algebraic expression. Let x represent the unknown number.

1. double a number

2. 8 less than a number

1. _____

2. _____

3. the quotient of –14 and a number

4. 21 more than 4 times a number

3. _____

4. _____

In Problems 5–7, choose a variable to represent one quantity. State what that quantity represents and then express the second quantity in terms of the first.

5. The Toronto Blue Jays scored 3 fewer runs than the Cleveland Indians.

5. _____

6. Beryl has $0.25 more than 3 times the amount Ralph has.

6. _____

7. There were 12,765 fans at a recent NBA game. Some held paid admission tickets and some held special promotion tickets.

7. _____

In Problems 8–11, translate each statement into an equation. Let x represent the unknown number. DO NOT SOLVE.

8. The sum of 43 and a number is –72.

9. 49 is 3 less than twice a number.

8. _____

9. _____

10. The quotient of a number and –6, decreased by 15, is 30.

11. Twice the sum of a number and 5 is the same as 7 more than the number.

10. _____

11. _____

In problems 12–15, translate the given information into an equation and solve.

12. Consecutive Integers The sum of three consecutive odd integers is 81. Find the numbers.

12. _____

13. Towers The tallest buildings in the world (those having the most stories) are the Sears Tower in Chicago, Illinois, and the Ryugyong Hotel in Pyongyang, North Korea. The Ryugyong Hotel has 5 fewer stories than the Sears Tower. The two buildings together have 215 stories. Find the number of stories in each building.

13. _____

14. Investments Jack and Diane have $40,000 to invest. Their financial advisor has recommended that they diversify by placing some of the money in stocks and some in bonds. Based upon current market conditions, he has recommended that the amount in bonds should equal two-thirds of the amount invested in stocks. How much should be invested in stocks? How much should be invested in bonds?

14. _____

15. Cellular Telephones You need a new cell phone for emergencies only. Company A charges $12 per month plus $0.10 per minute, while Company B charges $0.15 per minute with no monthly service charge. For how many minutes will the monthly cost be the same?

15. _____

Five-Minute Warm-Up 2.6
Problem Solving: Direct Translation Problems Involving Percent

1. Write as a decimal.

   (a) 62%

   (b) 1.75%

   1a. _____

   1b. _____

2. Write as a percent.

   a) 0.055

   (b) 1.5

   2a. _____

   2b. _____

3. Combine like terms: $p + 0.75p$

   3. _____

4. Solve: $x - 0.20x = 10.60$

   4. _____

Guided Practice 2.6
Problem Solving: Direct Translation Problems Involving Percent

Objective 1: Solve Direct Translation Problems Involving Percent

1. Percent men means "divided by 100". For instance, 20% means $\dfrac{20}{100}$ or $\dfrac{1}{5}$. Write each of these percents as fractions. Simplify, if possible.

    (a) 75%   **(b)** 100%   **(c)** $5\dfrac{1}{4}\%$ 1a. _____

                                                                     1b. _____

                                                                     1c. _____

2. To divide by 100, shift the decimal point 2 decimal places to the left. Write each of these percents as decimals.

    (a) 68%   **(b)** 2.5%   **(c)** 150%   2a. _____

                                                                      2b. _____

                                                                     2c. _____

3. If a question says, "what percent" and the solution to our equation is either a fraction or a decimal, we convert the result to percent by reversing the processes above. Write each of the following as a percent.

    (a) 0.08   **(b)** $\dfrac{7}{10}$ **(c)** 0.3   3a. _____

                                                                      3b. _____

                                                                     3c. _____

4. A number is 72% of 30. Find the number. *(See textbook Example 1)*
    (a) Translate the statement into an equation.   4a. _____

    (b) Solve the equation and answer the question.   4b. _____

5. The number 150 is what percent of 120? *(See textbook Example 2)*
    (a) Translate the statement into an equation.   5a. _____

    (b) Solve the equation and answer the question.   5b. _____

Objective 2: Model and Solve Direct Translation Problems from Business That Involve Percent

6. You just purchased a new computer. The price of the computer, including 7.5% sales tax, was $1343.75. How much did the computer cost before sales tax? *(See textbook Example 5)*

> **Step 1: Identify** We want to know the price of the computer before sales tax.
>
> **Step 2: Name** Let p represent the price of the computer before sales tax.

(a) Step 3: Translate If p is the price of the computer, what algebraic expression calculates the amount of sales tax on the computer?

_____

(b) Write the equation: the original price plus the sales tax equals the total price: _____

(c) Step 4: Solve Solve your equation from Step 3. _____

> **Step 5: Check** Does your answer seem reasonable? Have you satisfied the criteria in the question?

(d) Step 6: Answer Always be careful to answer the question that is being asked. _____

7. Suppose you just learned that the local sporting goods store is having an end-of-the-season sale on ski equipment. Everything has been marked down by 60%. The sale price on a pair of Black Diamond skis is $224. What was the original price? *(See textbook Example 6)*

> **Step 1: Identify** We are looking for the original price of the skis, which have been marked down by 60%, and we know that the sale price is $224.
>
> **Step 2: Name** Let p represent the original price of the skis.

(a) Step 3: Translate If p is the original price of the skis, what algebraic expression calculates the amount of discount?

_____

(b) Write the equation: the original price less the discount equals the sale price:

_____

(c) Step 4: Solve Solve your equation from Step 3. _____

> **Step 5: Check** Does your answer seem reasonable? Have you satisfied the criteria in the question?

(d) Step 6: Answer Always be careful to answer the question that is being asked. _____

Do the Math Exercises 2.6
Problem Solving: Direct Translation Problems Involving Percent

In Problems 1–8, find the unknown in each percent question.

1. What is 85% of 50?

2. 75% of 20 is what number?

1. _____

2. _____

3. What number is 150% of 9?

4. 40% of what number is 122?

3. _____

4. _____

5. 45% of what number is 900?

6. 11 is 5.5% of what number?

5. _____

6. _____

7. 4 is what percent of 25?

8. What percent of 16 is 12?

7. _____

8. _____

9. Sales Tax The sales tax in Franklin County, Ohio, is 5.75%.The total cost of purchasing a used Honda Civic, including sales tax, is $8460. Find the cost of the car before sales tax.

9. _____

10. Pay Raise MaryBeth works from home as a graphic designer. Recently she raised her hourly rate by 5% to cover increased costs. Her new hourly rate is $23.70. Find MaryBeth's previous hourly rate.

10. _____

11. Good Investment Perry just learned that his house increased in value by 4% over the past year. The value of the house is now $208,000. What was the value of the home one year ago?

11. _____

12. Hailstorm Toyota Town had a 15%-off sale on cars that had been damaged in a hailstorm. A new Toyota truck is on sale for $13,217.50. What was the price of the truck before the hailstorm discount?

12. _____

13. Business: Marking up the Price of Books A college bookstore marks up the price that it pays the publisher for a book by 35%. If the selling price of a book is $56.00, how much did the bookstore pay for the book?

13. _____

14. Vacation Package The Liberty Travel Agency advertised a 5-night vacation package in Jamaica for 30% off the regular price. The sale price of the package is $1139. To the nearest dollar, how much was the vacation package before the 30% off sale?

14. _____

15. Census Data Based on data obtained from the U.S. Census Bureau, 22% of the 113 million females aged 18 years or older have never married. How many females aged 18 years or older have never married?

15. _____

Five-Minute Warm-Up 2.7
Problem Solving: Geometry and Uniform Motion

1. Solve: $a + 2(a - 30) = 90$ 1. _____

2. Solve: $2(x + 5) + 2(3x - 8) = 22$ 2. _____

3. If x represents the unknown quantity, translate each of the following English phrases to an algebraic expression.

   (a) 12 more than twice an unknown number   3a. _____

   (b) 25 less than half of the width   3b. _____

   (c) double the difference of 45 and the measure of an angle   3c. _____

   (d) the sum of 15 and twice the cost   3d. _____

   (e) twice the sum of the speed and 30   3e. _____

4. If an 8-foot board is cut into two pieces and one piece has length p, express the length of the other piece in terms of p.   4. _____

Guided Practice 2.7
Problem Solving: Geometry and Uniform Motion

Objective 1: Set Up and Solve Complementary and Supplementary Angle Problems

1. Two angles are *complementary* if the sum of their measures is _____. Each angle is called the *complement* of the other.

2. Two angles are *supplementary* if the sum of their measures is _____. Each angle is called the *supplement* of the other.

3. Find the measures of two supplementary angles such that the measure of the smaller angle is 20° less than the measure of the larger angle. *(See textbook Example 1)*

   Step 1: Identify Since the angles are supplementary, we know the sum of their measures is 180°.

   Step 2: Name Let x represent the measure of the larger angle.

   (a) Step 3: Translate If x is the measure of the larger angle, what algebraic expression calculates the measure of the smaller angle?

   _____

   (b) Write the equation: the sum of the measures is equal to 180:   _____

   (c) Step 4: Solve your equation from Step 3.   _____

   Step 5: Check Does your answer seem reasonable? Have you satisfied the criteria in the question?

   (d) Step 6: Answer Always be careful to answer the question that is being asked. _____

Objective 2: Set Up and Solve Angles of a Triangle Problem

4. The sum of the measures of the three angles of a triangle is always _____.

5. The measure of the third angle of a triangle is 30° more than measure of the second angle of a triangle and the measure of the second angle is twice the measure the first angle. Find the measure of each angle of the triangle. *(See textbook Example 2)*

   Step 1: Identify Since these are angles of a triangle, we know the sum of their measures is 180°.

   Step 2: Name Let x represent the measure of the first angle.

   (a) Step 3: Translate If x is the measure of the first angle, what algebraic expression calculates the measure of the second angle?

   _____

5. **(b)** What algebraic expression calculates the measure of the third angle? _____

    (c) Write the equation: the sum of the measures is equal to 180: _____

    (d) **Step 4: Solve** your equation from Step 3. _____

        Step 5: Check Does your answer seem reasonable? Have you satisfied the criteria in the question?

    (e) **Step 6: Answer** Always be careful to answer the question that is being asked. _____

Objective 3: Use Geometry Formulas to Solve Problems

6. Tablecloth Mary is making a rectangular tablecloth that is 20 inches longer than it is wide. If she has 200 inches of trim that will just fit around the edge of the tablecloth, find the dimensions of the tablecloth. *(See textbook Example 3)*

    (a) If x is width of the tablecloth, what algebraic expression represents the length? _____

    (b) What formula from geometry is needed to solve this problem? _____

    (c) Write an equation that models the information in the problem: _____

    (d) Solve your equation from (c) and answer the question: _____

Objective 4: Set Up and Solve Uniform Motion Problems

7. Boats Two boats are traveling toward the same port from opposite directions. They are 50 miles apart and one boat is traveling 4 mph faster than the other. If the boats both reach the port in 2 hours and 30 minutes, find the speed for each boat. *(See textbook Example 6)*

    (a) What do you know about the distances traveled by the two boats? _____

    (b) Complete the following table.

	Rate, mph	Time, hours	Distance, miles
Slower boat			
Faster boat			
Total			

    (c) Write and solve an equation that models the distance traveled by the boats.

    (d) Answer the question. _____

Do the Math Exercises 2.7
Problem Solving: Geometry and Uniform Motion

Find the value of x and then identify the measure of each of the angles.

1. Find two complementary angles such that the measure of the first angle is 25° less than the measure of the second.

   1. _____

Find the measures of the angles of the triangle.

2. In a triangle, the second angle measures four times the first. The measure of the third angle is 18° more than the second. Find the measures of the three angles.

   2. _____

Use a formula from geometry to solve for the unknown quantity.

3. The width of a rectangle is 10 m less than half of the length. If the perimeter is 52 meters, find the length of each side of the rectangle.

   3. _____

Solve the uniform motion problem. Set up the uniform motion problem following steps 4a–4d.

4. Two trains leave Albuquerque, traveling the same direction on parallel tracks. One train is traveling at 72 mph and the other is traveling at 66 mph. How long before they are 45 miles apart?

   (a) Write an algebraic expression for the distance traveled by the faster train.

   4a. _____

   (b) Write an algebraic expression for the distance traveled by the slower train.

   4b. _____

   (c) Write an algebraic expression for the difference in distance between the two.

   4c. _____

   (d) Write an equation to answer the question.

   4d. _____

Do the Math Exercises 2.7

Fill in the table from the information given. Then write the equation that will solve the problem. DO NOT SOLVE.

5. A 580-mile trip in a small plane took a total of 5 hours. The first two hours were flown   5. _____
at one rate and then the plane encountered a headwind and was slowed by 10 mph.

	Rate	•	Time	=	Distance
Beginning of trip					
Rest of the trip					
Total					

Write an equation that will solve each problem. Then solve.

6. **Garden** The length of a rectangular garden is 9 feet. If 26 feet of fencing are required to   6. _____
fence the garden, find the width of the garden.

7. **Buying Fertilizer** Melinda has to buy fertilizer for a flower garden in the shape of a   7. _____
right triangle. If the area of the garden is 54 square feet and the base of the garden
measures 9 feet, find the height of the triangular garden.

8. **Table** The Jacksons are having a custom rectangular table made for a small dining area.   8. _____
The length of the table is 18 inches more than the width, and the perimeter is 180 inches.
Find the length and width of the table.

9. **Cyclists** Two cyclists leave a city at the same time, one going east and the other going   9. _____
west. The west-bound cyclist bikes at 4 mph faster than the east-bound cyclist. After 5
hours they are 200 miles apart. How fast is the east-bound cyclist riding?

10. **River Trip** Max lives on a river, 30 miles from town. Max travels downstream (with the   10. _____
current) at 20 mph. Returning upstream (against the current) his rate is 12 mph. If the
total trip to town and back took 4 hours, how long did he spend returning from town?

Sullivan/Struve/Mazzarella, *Elementary Algebra,* 2e

Five-Minute Warm-Up 2.8
Solving Linear Inequalities in One Variable

In Problems 1 – 6, replace the question mark by <, >, or = to make the statement true.

1. -0.2 ? 0

2. -5 ? -4.5

3. $\dfrac{7}{12}$? $\dfrac{9}{16}$

4. 0.625 ? $\dfrac{5}{8}$

5. $\dfrac{-13}{8}$? $\dfrac{-11}{4}$

6. 0.003 ? 0.0031

1. _____

2. _____

3. _____

4. _____

5. _____

6. _____

7. Plot the set of points $\left\{ -3.5, \dfrac{3}{2}, -\dfrac{2}{3}, 3 \right\}$ on the number line.

Guided Practice 2.8
Solving Linear Inequalities in One Variable

Objective 1: Graph Inequalities on a Real Number Line

1. When representing an inequality on a number line, we graph the inequality using a(n) _____

if the endpoint in included and a(n) _____ if the endpoint is not included.

2. Graph $x > -2$ on a real number line. *(See textbook Example 1)*

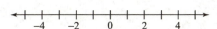

Objective 2: Use Interval Notation

3. Write the inequality in interval notation. *(See textbook Example 2)*

   (a) $x < -5$ **(b)** $x \geq 10$ 3a. _____

                                                          3b. _____

Objective 3: Solve Linear Inequalities Using Properties of Inequalities

4. *True or False:* The Addition Property of Inequality states that the direction of the
inequality does not change regardless of the quantity that is added to each side of the   4. _____
inequality.

5. *True or False:* When multiplying both sides of an inequality by a negative number, we   5. _____
reverse the direction of the inequality.

6. *True or False:* When multiplying both sides of an inequality by a positive number, we   6. _____
reverse the direction of the inequality.

7. Solve the inequality $3x + 10 > 1$. Graph the solution set. *(See textbook Examples 3 and 4)*

Step 1: Get the expression containing the variable on the left side of the inequality.	This is already done.	$3x + 10 > 1$
Step 2: Isolate the variable x on the left side.	Subtract 10 from both sides:	**(a)** $3x + 10$ _____ > 1 _____
	Simplify:	**(b)** _____
	Divide both sides by 3 and simplify:	**(c)** _____
	Write the answer in set-builder notation:	**(d)** _____
	Write the answer using interval notation:	**(e)** _____
	Graph the solution set:	**(f)**

8. When multiplying or dividing both sides of an inequality by a negative real number, remember to reverse the direction of the inequality symbol. Solve the inequality and state the solution using interval notation. *(See textbook Example 6)*

(a) $-6x < -72$ (b) $-\dfrac{2}{5}x \geq 10$ 8a. _____

8b. _____

9. Solve the inequality $\dfrac{1}{6}(4x - 5) \leq \dfrac{1}{2}(2x + 3)$. *(See textbook Examples 7 and 8)*

Step 1: To rewrite the inequality as an equivalent inequality without fractions, we multiply both sides of the inequality by the LCD. Then remove parentheses.	Multiply both sides by the LCD:	$\dfrac{1}{6}(4x - 5) \leq \dfrac{1}{2}(2x + 3)$ (a) ___ $\bullet \dfrac{1}{6}(4x - 5) \leq$ ___ $\bullet \dfrac{1}{2}(2x + 3)$
	Divide out common factors:	**(b)** _____
	Use the Distributive Property:	**(c)** _____
Step 2: Combine like terms on each side of the inequality.		This is already done.
Step 3: Get the variable expressions on the left side of the inequality and the constants on the right side.	Subtract 6x from both sides:	**(d)** _____
	Add 5 to both sides:	**(e)** _____
Step 4: Get the coefficient of the variable to be one.	Divide both sides by −2: Remember to reverse the inequality symbol!	**(f)** _____
	Write the answer in set-builder notation:	**(g)** _____
	Write the answer using interval notation:	**(h)** _____

Objective 4: Model Inequality Problems

10. Write the appropriate inequality symbol for each phrase.

Phrase	Inequality Symbol	Phrase	Inequality Symbol
(a) At least	**(a)** _____	**(b)** No less than	**(b)** _____
(c) More than	**(c)** _____	**(d)** Greater than	**(d)** _____
(e) No more than	**(e)** _____	**(f)** At most	**(f)** _____
(g) Fewer than	**(g)** _____	**(h)** Less than	**(h)** _____

Do the Math Exercises 2.8
Solving Linear Inequalities in One Variable

In Problems 1–4, graph each inequality on a number line, and write each inequality in interval notation.

1. $n > 5$

2. $x \le 6$

1. _____

2. _____

3. $x \ge -2$

4. $y < -3$

3. _____

4. _____

In Problems 5–6, use interval notation to express the inequality shown in each graph.

5.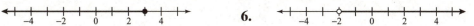

6.

5. _____

6. _____

In Problems 7–8, fill in the blank with the correct symbol. State which property of inequality is being utilized.

7. If $3x - 2 > 7$, then $3x$ _____ 9

8. If $\dfrac{5}{3}x \ge -10$, then x _____ -6

7. _____

8. _____

In Problems 9–17, solve the inequality and express the solution set in set-builder notation and interval notation Graph the solution set on a real number line.

9. $x - 2 < 1$

10. $4x > 12$

9. _____

10. _____

11. $-7x \geq 28$

12. $2x + 5 > 1$

11. _____

12. _____

13. $2x - 2 \geq 3 + x$

14. $2 - 3x \leq 5$

13. _____

14. _____

15. $-3(1-x) > x + 8$

16. $2y - 5 + y < 3(y-2)$

15. _____

16. _____

17. $3(p+1) - p \geq 2(p+1)$

17. _____

In Problems 18–19, write the given statement using inequality symbols. Let x represent the unknown quantity.

18. The cost of a new lawnmower is at least $250.

18. _____

19. There are fewer than 25 students in your math class on any given day.

19. _____

20. Truck Rental A truck can be rented from Acme Truck Rental for $80 per week plus $0.28 per mile. How many miles can be driven if you have at most $100 to spend on truck rental?

20. _____

Sullivan/Struve/Mazzarella, *Elementary Algebra*, 2e

Five-Minute Warm-Up 3.1
The Rectangular Coordinate System and Equations in Two Variables

1. Plot the following points on the real number line: $-1, \dfrac{3}{2}, 3, -2.5$

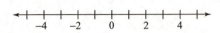

2. Evaluate: $-2x - 3$ 2a. _____

   (a) for $x = -5$ **(b)** for $x = 0$

                                                                       2b. _____

3. Evaluate: $7x + 3y$ 3a. _____

   (a) for $x = -1, y = -6$ **(b)** for $x = 3, y = -4$

                                                                         3b. _____

4. Solve: $-2x + 9 = -1$ 4. _____

5. Solve: $4(2x - 1) - 6x = 2 - 5(x - 3)$ 5. _____

Guided Practice 3.1
The Rectangular Coordinate System and Equations in Two Variables

Objective 1: Plot Points in the Rectangular Coordinate System

1. Label each quadrant and axis in the rectangular or Cartesian coordinate system.

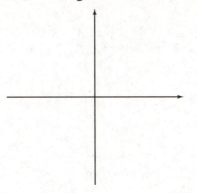

2. What name do we give the ordered pair $(0, 0)$? _____

In Problem 3, circle one answer for each underlined choice.

3. To plot the ordered pair $(-3, 2)$, you would move 3 units <u>up, down, left or right</u> from the origin?

4. Identify the coordinates of each point labeled below. (*See textbook Example 2*)

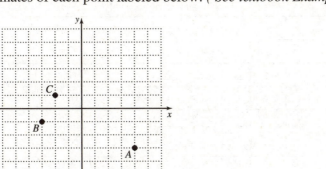

4.

A _____

B _____

C _____

Objective 2: Determine If an Ordered Pair Satisfies an Equation

5. Determine which, if any, of the following ordered pairs satisfy the equation $2x - y = -5$.
(See textbook Example 3)

   (a) $(1, -3)$ **(b)** $(-5, -5)$ 5a. _____

                                                                          5b. _____

6. Find an ordered pair that satisfies the equation $4x + 2y = 7$. *(See textbook Example 4)*

Step 1: Choose any value for one of the variables in the equation.	You may choose any value of x or y that you wish. In this example, we will let $x = 3$.	
Step 2: Substitute the value of the variable chosen in Step 1 into the equation and then use the techniques learned in Chapter 2 to solve for the remaining variable.	Substitute 3 for x in the equation $4x + 2y = 7$ and then solve for y:	$4x + 2y = 7$ $4(3) + 2y = 7$ **(a)** $2y = 7 - 4(3) =$ _____
	Divide both sides by 2 and simplify:	**(b)** $y =$ _____
	Write the complete solution as an ordered pair:	**(c)** $(x, y) =$ _____

Objective 3: Create a Table of Values That Satisfy an Equation

7. To create a table of values that satisfy an equation, choose any value for one of the variables in the equation. Substitute the value of the variable into the equation and then use the techniques learned in Chapter 2 to solve for the remaining variable. Complete the table of values for $y = 3x - 1$. *(See textbook Example 5)*

	x	y	(x, y)
(a)	-2	$3(\underline{\ \ }) - 1 = \underline{\ \ }$	$(-2, \underline{\ \ })$
(b)	-1	$3(\underline{\ \ }) - 1 = \underline{\ \ }$	$(-1, \underline{\ \ })$
(c)	0	$3(\underline{\ \ }) - 1 = \underline{\ \ }$	$(0, \underline{\ \ })$
(d)	1	$3(\underline{\ \ }) - 1 = \underline{\ \ }$	$(1, \underline{\ \ })$
(e)	2	$3(\underline{\ \ }) - 1 = \underline{\ \ }$	$(2, \underline{\ \ })$

Do the Math Exercises 3.1
The Rectangular Coordinate System and Equations in Two Variables

In Problems 1–2, plot the following ordered pairs in the rectangular coordinate system. Tell which quadrant each point lies in or state that the point lies on the x-axis or y-axis.

1. $P(-3, -2)$; $Q(2, -4)$; $R(4, 3)$; $S(-1, 4)$; $T(-2, -4)$; $U(3, -3)$

2. $P\left(\dfrac{3}{2}, -2\right)$; $Q\left(0, \dfrac{5}{2}\right)$; $R\left(-\dfrac{9}{2}, 0\right)$; $S(0, 0)$; $T\left(-\dfrac{3}{2}, -\dfrac{9}{2}\right)$; $U\left(3, \dfrac{1}{2}\right)$ $V\left(\dfrac{5}{2}, -\dfrac{7}{2}\right)$

1. _____

2. _____

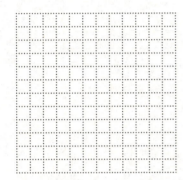   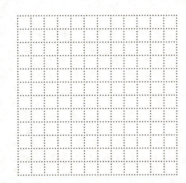

Identify the coordinates of each point labeled in the figure. Name the quadrant in which each point lies or state that the point lies on the x- or y-axis.

3.

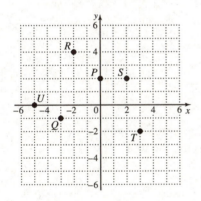

3. _____

In Problems 4–5, determine whether or not the ordered pair satisfies the equation.

4. $y = 2x - 3$ $A(-1, -5)$ $B(4, -5)$ $C(-2, -7)$

5. $5x - y = 12$ $A(2, 0)$ $B(0, 12)$ $C(-2, -22)$

4. _____

5. _____

6. Find an ordered pair that satisfies the equation $x + y = 7$ by letting $x = 2$.

7. Find an ordered pair that satisfies the equation $5x - 3y = 11$ by letting $y = 3$.

6. _____

7. _____

In Problems 8–9, use the equation to complete the table. Use the table to list some of the ordered pairs that satisfy the equation.

8. $y = 4x - 5$

x	y	(x, y)
−3		
1		
2		

9. $3x + 4y = 2$

x	y	(x, y)
−2		
2		
4		

In Problems 10–11, for each equation find the missing value in the ordered pair.

10. $y = 5x - 4$ $A(-1, \underline{\quad})$ $B(\underline{\quad}, 31)$

$C\left(-\dfrac{2}{5}, \underline{\quad}\right)$

11. $\dfrac{1}{3}x + 2y = -1$ $A(-4, \underline{\quad})$ $B\left(\underline{\quad}, -\dfrac{3}{4}\right)$

$C(0, \underline{\quad})$

10. _____

11. _____

12. Taxi Ride The cost to take a taxi is $1.70 plus $2.00 per mile for each mile driven. The total cost, C, is given by the equation $C = 1.7 + 2m$ where m represents the total miles driven.

(a) How much will it cost to take a taxi 5 miles?

12a. _____

(b) How much will it cost to take a taxi 20 miles?

12b. _____

(c) If you spent $32.70 on cab fare, how far was your trip?

12c. _____

(d) If (m, C) represents any ordered pair that satisfies $C = 1.7 + 2m$, interpret the meaning of (14, 29.7) in the context of this problem.

12d. _____

Five-Minute Warm-Up 3.2
Graphing Equations in Two Variables

1. Solve: $-5x = 45$

    1. _____

2. Solve: $12y = -40$

    2. _____

3. Solve: $5y + 12 = 2$

    3. _____

4. Evaluate: $-2x + 8y$
   (a) for $x = -3$, $y = 2$ **(b)** for $x = 10$, $y = 4$

    4a. _____

    4b. _____

5. Solve using the given information: $y = \dfrac{3}{2}x + 5$
   (a) If $x = -4$, find y.   **(b)** If $y = -1$, find x.

    5a. _____

    5b. _____

6. Plot the ordered pairs: $(0, -3)$; $(2, 0)$; $(-3, 2)$; $(1, -4)$; $(-2, -1)$

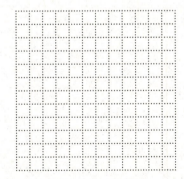

Guided Practice 3.2
Graphing Equations in Two Variables

Objective 1: Graph a Line by Plotting Points

1. Point-plotting is one method we use to graph a line. Find several ordered pairs that satisfy the equation. Plot the points in a rectangular coordinate system and then connect the points in a smooth curve or line.

Graph the equation $y = -2x - 3$ using the point-plotting method. *(See textbook Example 1)*

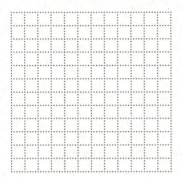

2. When a linear equation in two variables is written in the form $Ax + By = C$ where A, B, and C are real

numbers and both A and B cannot be 0, we say the linear equation is written in _____ form.

Objective 2: Graph a Line Using Intercepts

3. To find the x-intercept(s), if any, of the graph of an equation, let _____ in the equation and solve for x.

4. To find the y-intercept(s), if any, of the graph of an equation, let _____ in the equation and solve for y.

5. Graph the linear equation $3x - 2y = -6$ by finding its intercepts. *(See textbook Example 6)*

    (a) Identify the x-intercept: _____   **(b)** Identify the y-intercept: _____

    (c) Draw the axes. Plot the intercepts and connect the points in a straight line to obtain the graph.

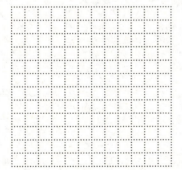

Objective 3: Graph Vertical and Horizontal Lines

6. A vertical line is given by an equation of the form $x = a$ where a is the x-intercept.

Graph the equation $x = -2$. *(See textbook Example 9)*

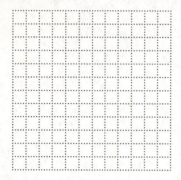

7. A horizontal line is given by an equation of the form $y = b$ where b is the y-intercept.

Graph the equation $y = 3$. *(See textbook Example 10)*

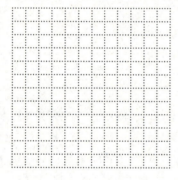

Do the Math Exercises 3.2
Graphing Equations in Two Variables

In Problems 1–2, determine whether or not the equation is a linear equation in two variables.

1. $y^2 = 2x + 3$

2. $y - 2x = 9$

1. _____

2. _____

In Problems 3–4, graph each linear equation using the point-plotting method.

3. $y = -3x - 1$

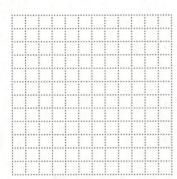

4. $y = x - 6$

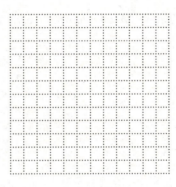

In Problems 5–8, find the intercepts of each graph.

5.

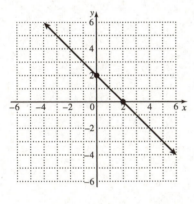

6.

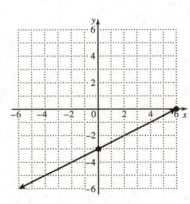

5. _____

6. _____

7.

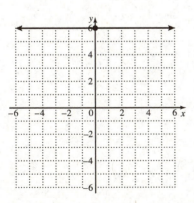

8.

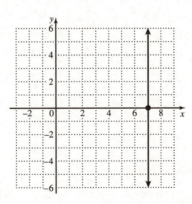

7. _____

8. _____

In Problems 9–10, find the intercepts of each equation.

9. $3x - 5y = 30$

10. $\dfrac{x}{2} - \dfrac{y}{8} = 1$

9. _____

10. _____

In Problems 11–12, graph each linear equation by finding its intercepts.

11. $3x - 5y = 15$

12. $\dfrac{1}{3}x - y = 0$

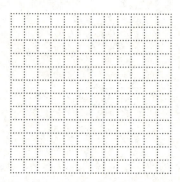

In Problems 13–14, graph each horizontal or vertical line.

13. $x = -7$

14. $y = 2$

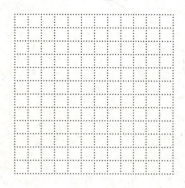

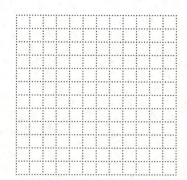

Sullivan/Struve/Mazzarella, *Elementary Algebra*, 2e

Five-Minute Warm-Up 3.3
Slope

In Problems 1 – 7, simplify each expression.

1. $\dfrac{9-3}{15-3}$

2. $\dfrac{6-15}{2-5}$

1. _____

2. _____

3. $\dfrac{-2-8}{4-(-2)}$

4. $\dfrac{-5-(-9)}{3-5}$

3. _____

4. _____

5. $\dfrac{5-12}{-6-(-6)}$

6. $\dfrac{-18-(-18)}{4-(-4)}$

5. _____

6. _____

7. $\dfrac{\frac{1}{2}-\left(-\frac{3}{4}\right)}{-\frac{2}{3}-\frac{5}{6}}$

7. _____

Guided Practice 3.3
Slope

Objective 1: Find the Slope of a Line Given Two Points

1. The ratio of the rise (vertical change) to the run (horizontal change) is called the _____ of the line.

2. The slope of a line which passes through the ordered pairs (x_1, y_1) and (x_2, y_2) is given by the formula:

$$m = \frac{\overline{}}{\underline{}}$$

3. Find the slope of the line containing the points $(-1, 7)$ and $(3, -13)$. *(See textbook Example 1)*

3. _____

Objective 2: Find the Slope of vertical and Horizontal Lines

4. Calculate the slope of the line through the points $(-8, 3)$ and $(-8, -1)$. *(See textbook Example 3)*

4. _____

5. Is the graph of the line in #4 vertical or horizontal?

5. _____

6. Calculate the slope of the line through the points $(-1, 5)$ and $(4, 5)$. *(See textbook Example 4)*

6. _____

7. Is the graph of the line in #6 vertical or horizontal?

7. _____

8. *Properties of Slope* Describe the graph of the line with the following slope:

(a) Positive _____   (b) Negative _____

(c) Zero _____   (d) Undefined _____

Objective 3: *Graph a Line Using Its Slope and a Point on the Line*

9. Draw a graph of the line that contains the point $(-1, 3)$ and has slope $-\dfrac{3}{4}$. *(See textbook Example 5)*

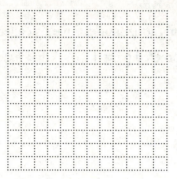

Sullivan/Struve/Mazzarella, *Elementary Algebra,* 2e

Do the Math Exercises 3.3
Slope

In Problems 1–2, find the slope of the line whose graph is given.

1.

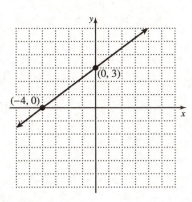

2.

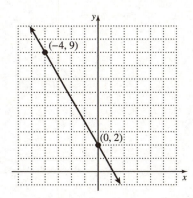

1. _____

2. _____

In Problems 3–4, (a) plot the points in a rectangular coordinate system, (b) draw a line through the points, (c) and find and interpret the slope of the line.

3. (2, 6) and (–2, –4)

4. (4, –5) and (–2, –4)

3c. _____

4c. _____

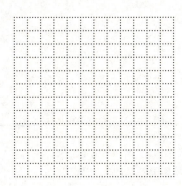

In Problems 5–6, find the slope of the line whose graph is given.

5.

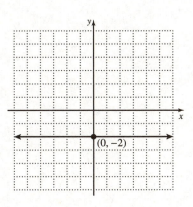

6.

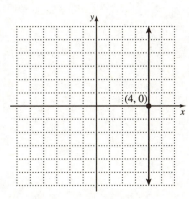

5. _____

6. _____

In Problems 7–8, draw a graph of the line that contains the given point and has the given slope.

7. $(3, -1)$; $m = -1$

8. $(-2, -3)$; $m = \dfrac{5}{2}$

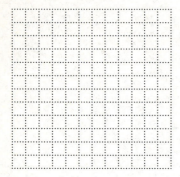

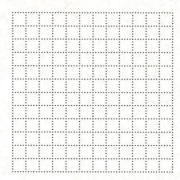

9. Roof Pitch A canopy is set up on the football field. On the 45-yard line, the height of the canopy is 68 inches. The peak of the canopy is at the 50-yard marker where the height is 84 inches. What is the pitch of the roof of the canopy?

9.

10. Earning Potential On average, a person who graduates from high school can expect to have lifetime earnings of 1.2 million dollars. It takes four years to earn a bachelor's degree, but the lifetime earnings will increase to 2.1 million dollars. Use the ordered pairs (0, 1.2 million) and (4, 2.1 million) to find and interpret the slope of the line representing the increase in earnings due to finishing college.

10. _____

Five-Minute Warm-Up 3.4
Slope-Intercept Form of a Line

1. Solve $7x + 5y = -35$ for y.   1. _____

2. Solve $3x - 8y = 12$ for y.   2. _____

3. Solve: $15 = 3 - 9x$ for x.   3. _____

4. Identify the following information using: $y = -\dfrac{5}{3}x - \dfrac{1}{2}$ 4a. _____

(a) the coefficient of x **(b)** the constant

4b. _____

Guided Practice 3.4
Slope-Intercept Form of a Line

Objective 1: Use the Slope-Intercept Form to Identify the Slope and y-Intercept of a Line

1. The *Slope-Intercept Form* of a line is an equation written in the form
$y = mx + b$, where the slope of the line is ____, the coefficient of *x*, and the
y-intercept is ____, the constant.

1. _____

_____

2. Find the slope and *y*-intercept of the line whose equation is $y = -\dfrac{1}{2}x + 6$.
(See textbook Example 1)

2. slope = _____

y-intercept = _____

3. Find the slope and *y*-intercept of the line whose equation is $6x + 3y = 15$.
(See textbook Example 2)

3. slope = _____

y-intercept = _____

Objective 2: Graph a Line Whose Equation Is in Slope-Intercept Form

4. Graph the line $y = \dfrac{2}{5}x - 1$ using the slope and *y*-intercept. *(See textbook Example 4)*

(a) What is the slope? _____

(b) What is the *y*-intercept? _____

(c) Graph the line.

Objective 3: Graph a Line Whose Equation Is in the Form Ax + By = C

5. Graph the line $2x + 3y = 6$ using the slope and *y*-intercept. *(See textbook Example 5)*

(a) Write the equation in slope-intercept form.

_____

(b) What is the slope? _____

(c) What is the *y*-intercept? _____

(d) Graph the line.

Objective 4: *Find the Equation of a Line Given Its Slope and y-Intercept*

6. Find the equation of a line whose slope is $-\dfrac{1}{4}$ and whose *y*-intercept is 3. *(See textbook Example 6)*

6. _____

7. The cost of renting a particular car for a day is $22 plus $0.07 per mile. Find the following. *(See textbook Example 8)*

(a) Write a linear equation that relates the daily cost of renting the car, *y*, to the number of miles driven in a day.

7a. _____

(b) What is the cost of driving this car 100 miles?

7b. _____

Sullivan/Struve/Mazzarella, *Elementary Algebra*, 2e

Do the Math Exercises 3.4
Slope-Intercept Form of a Line

In Problems 1–4, find the slope and y-intercept of the line whose equation is given.

1. $y = -6x + 2$

2. $y = -x - 12$

3. $6x - 8y = -24$

4. $10x + 6y = 24$

1. _____

2. _____

3. _____

4. _____

In Problems 5–6, use the slope and y-intercept to graph each line whose equation is given.

5. $y = -4x - 1$

6. $y = \dfrac{4}{3}x - 3$

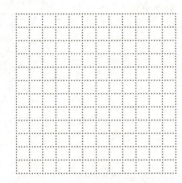
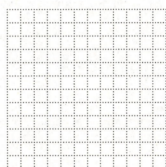

In Problems 7–8, graph each line using the slope and y-intercept.

7. $x - 2y = -4$

8. $4x + 3y = -6$

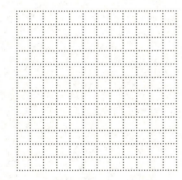
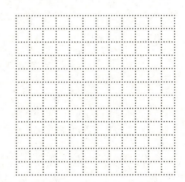

In Problems 9–14, find the equation of the line with the given slope and intercept.

9. slope is 1; *y*-intercept is 10

10. slope is $\dfrac{4}{7}$; *y*-intercept is -9

9. _____

10. _____

11. slope is $\dfrac{1}{4}$; *y*-intercept is $\dfrac{3}{8}$

12. slope is 0; *y*-intercept is –2

11. _____

12. _____

13. slope is undefined; *x*-intercept is 4

14. slope is –3; *y*-intercept is 0

13. _____

14. _____

15. Counting Calories According to a 1989 National Academy of Sciences Report, the recommended daily intake of calories for males between the ages of 7 and 15 can be calculated by the equation $y = 125x + 1125$, where *x* represents the boy's age and *y* represents the recommended calorie intake.

(a) What is the recommended caloric intake for a 12-year-old boy?

15a. _____

(b) What is the age of a boy whose recommended caloric intake is 2250 calories?

15b. _____

(c) Interpret the slope of $y = 125x + 1125$.

15c. _____

(d) Why would this equation not be accurate for a 3-year-old male?

15d. _____

(e) Graph the equation in a rectangular coordinate system. Label the axes appropriately.

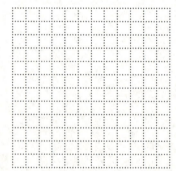

Sullivan/Struve/Mazzarella, *Elementary Algebra*, 2e

Five-Minute Warm-Up 3.5
Point-Slope Form of a Line

1. Solve $y - (-2) = -4(x - 3)$ for y.

1. _____

2. Solve $9x - 3y = -12$ for y.

2. _____

3. Evaluate: $\dfrac{8 - 5}{2 - 11}$

3. _____

4. Evaluate: $\dfrac{0.4 - (-0.2)}{9.9 - 10.2}$

4. _____

5. Use the Distributive Property to simplify: $-\dfrac{7}{4}(4x - 12)$

5. _____

6. Find the slope of the line that contains $(-7, 10)$ and $(-3, 6)$.

6. _____

Guided Practice 3.5
Point-Slope Form of a Line

Objective 1: Find the Equation of a Line Given a Point and a Slope

1. What equation do we use to write the *Point-Slope Form* of a line with slope m and containing (x_1, y_1) ?

_____

2. Find the equation of a line whose slope is 9 and that contains the point $(-2, -1)$. *(See textbook Example 1)*

    (a) Identify values: $m = $ _____; $x_1 = $ _____; $y_1 = $ _____

    (b) Point-slope form of a line: _____

    (c) Substitute the values from (a) into the equation from (b): _____

    (d) Simplify: _____

    (e) Use the Distributive Property: _____

    (f) Simplify and write the equation in slope-intercept form: _____

3. Find the equation of a line whose slope is $-\dfrac{5}{6}$ and that contains the point $(0, 12)$. *(See textbook Example 2)*

    (a) Identify values: $m = $ _____; $x_1 = $ _____; $y_1 = $ _____

    (b) Point-slope form of a line: _____

    (c) Substitute the values from (a) into the equation from (b): _____

    (d) Simplify and use the Distributive Property: _____

4. Find the equation a horizontal line that contains the point $(5, -11)$. *(See textbook Example 3)*

    (a) Identify values: $m = $ _____; $x_1 = $ _____; $y_1 = $ _____

    (b) Point-slope form of a line: _____

    (c) Substitute the values from (a) into the equation from (b):_____

    (d) Simplify: _____

                                          _____

Objective 2: Find the Equation of a Line Given Two Points

5. To find the equation of a line given two points, the first step is to find _____.

6. Find the equation of the line through the points $(1, 5)$ and $(-1, -1)$. If possible, write the equation in slope-intercept form. *(See textbook Example 4)*

Step 1: Find the slope of the line containing the points.

Formula for the slope of a line through two points: **(a)** _____

Identify values: **(b)** $x_1 =$ _____; $y_1 =$ _____

$x_2 =$ _____; $y_2 =$ _____

Substitute values into (a) and simplify: **(c)** _____

Step 2: Substitute the slope found in Step 1 and either point into the point-slope form of a line to find the equation.

Point-slope form of a line: **(d)** _____

Identify values: **(e)** $m =$ _____; $x_1 =$ _____; $y_1 =$ _____

Substitute values into (e): **(f)** _____

Step 3: Solve the equation for y.

Distribute the 3: **(g)** _____

Add _____ to both sides: **(h)** _____

Is the equation in slope-intercept form? **(i)** _____

Try using (x_2, y_2) in (d). Does this change the equation of the line found in (i)? **(j)** _____

Identify the slope: **(k)** _____

Identify the y-intercept: **(l)** _____

7. Find the equation of a line through the points $(-5, -1)$ and $(-5, 3)$. *(See textbook Example 5)*

(a) Find the slope of the line: _____

(b) What do we know about the graph of this line? _____

(c) No matter what value of y we choose, the x-coordinate of the point on the line will be _____

(d) Write the equation of the line: _____

Do the Math Exercises 3.5
Point-Slope Form of a Line

In Problems 1–4, find the equation of the line that contains the given point and with the given slope. Write the equation in slope-intercept form and graph the line.

1. $(4, 1)$; slope $= 6$

2. $(6, -3)$; slope $= -5$

1. _____

2. _____

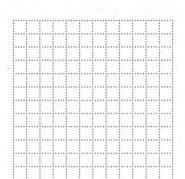

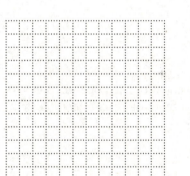

3. $(-8, 2)$; slope $= -\dfrac{1}{2}$

4. $(-7, -1)$; slope $= 0$

3. _____

4. _____

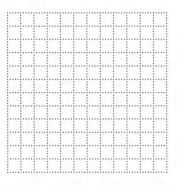

In Problems 5–6, find the equation of the line that contains the given point and satisfies the given information. Write the equation in slope-intercept form, if possible.

5. Horizontal line that contains $(-6, -1)$

6. Vertical line that contains $(4, -3)$

5. _____

6. _____

In Problems 7–10, find the equation of the line that contains the given points. Write the equation in slope-intercept form, if possible.

7. $(-2, 4)$ and $(2, -2)$

8. $(4, 18)$ and $(-1, 3)$

7. _____

8. _____

9. (–6, 5) and (7, 5)   **10.** (–3, 8) and (–3, 1)   9.  _____

                                                                                     10. _____

11. Retirement Plans Based on the retirement plan available by his employer, Kei knows that if he retires after 20 years, his monthly retirement income will be $3150. If he retires after 30 years, his monthly income increases to $3600. Let x represent the number of years of service and y represent the monthly retirement income.

(a) Interpret the meaning of the point (30, 3600) in the context of this problem.   11a. _____

(b) Plot the ordered pairs (20, 3150) and (30, 3600) in a rectangular coordinate system and graph the line through the points.

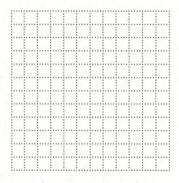

(c) Find the linear equation, in slope-intercept form, that relates the monthly retirement income, y, to the number of years of service, x.   11c. _____

(d) Use the equation found in part (c) to find the monthly income for 15 years of service. 11d. _____

(e) Interpret the slope.   11e. _____

Five-Minute Warm-Up 3.6
Parallel and Perpendicular Lines

1. Determine the reciprocal of -5.

   1. _____

2. Determine the reciprocal of $\frac{1}{3}$.

   2. _____

3. Determine the reciprocal of $-\frac{9}{4}$.

   3. _____

4. Solve $-4x - y = 12$ for y.

   4. _____

5. Identify the following information using: $-x + 4y = -2$

   (a) the slope of the line   (b) the y-intercept of the line

   5a. _____

   5b. _____

6. Find the equation of the line that has slope $-\frac{3}{2}$ and contains the point $(5, -2)$.

   6. _____

Guided Practice 3.6
Parallel and Perpendicular Lines

Objective 1: Determine Whether Two Lines Are Parallel

1. In your own words, write a definition of parallel lines.

2. Two lines are parallel if and only if they have the same _____ and different _____.

3. If $L_1 : y = m_1 x + b_1$ and $L_2 : y = m_2 x + b_2$ and L_1 is parallel to L_2, then m_1 _____ m_2 and b_1 _____ b_2.

4. Determine whether the line $4x - y = 1$ is parallel to $8x - 2y = -10$. *(See textbook Example 2)*

To find the slope and y-intercept, we solve each equation for y so that each is in slope-intercept form.

(a) Equation 1, $4x - y = 1$, written in slope-intercept form: _____

(b) Determine the slope and y-intercept of the line in equation 1:   slope: _____; y-intercept: _____

(c) Equation 2, $8x - 2y = -10$, written in slope-intercept form: _____

(d) Determine the slope and y-intercept of the line in equation 2:   slope: _____; y-intercept: _____

(e) Conclusion: _____

Objective 2: Find the Equation of a Line Parallel to a Given Line

5. Find an equation for the line that is parallel to $5x - 2y = -1$ and contains the point $(-3, -4)$.
(See textbook Example 3)

Step 1: Find the slope of the given line by putting the equation in slope-intercept form.

(a)_____

Step 2: Use the point-slope form of a line with the given point and the slope found in Step 1 to find the equation of the parallel line.

Slope of the line found in Step 1:   **(b)**_____

Slope of the parallel line: **(c)**_____

Identify values: **(d)** $m =$ ____; $x_1 =$ ____; $y_1 =$ _____

Point-slope form of a line: **(e)**_____

Substitute values into (e):   **(f)**_____

Step 3: Put the equation in slope-intercept form by solving for y.

Solve for y: **(g)**_____

Objective 3: Determine Whether Two Lines are Perpendicular

6. In your own words, write a definition of perpendicular lines.

7. Two lines are perpendicular if and only if the product of their slopes is _____.

8. If $L_1 : y = m_1 x + b_1$ and $L_2 : y = m_2 x + b_2$ and L_1 is perpendicular to L_2, then $m_1 \cdot m_2 =$ ___ or $m_1 =$ ___.

9. Determine whether the line $y = \dfrac{3}{4}x + 6$ is perpendicular to $y = \dfrac{4}{3}x - 2$. *(See textbook Example 6)*

    (a) Determine the slope and *y*-intercept of the line in equation 1:   slope: _____; *y*-intercept: _____

    (b) Determine the slope and *y*-intercept of the line in equation 2:   slope: _____; *y*-intercept: _____

    (c) What is the product of the slopes of the two lines?   _____

    (d) Conclusion: _____

10. Find an equation of the line that is perpendicular to the line $-2x + y = 7$ and contains the point $(3, -12)$. Write the equation in slope-intercept form. *(See textbook Example 8)*

Step 1: Find the slope of the given line by putting the equation in slope-intercept form.		**(a)**_____
Step 2: Find the slope of the perpendicular line.	Slope of the line found in Step 1:	**(b)**_____
	Slope of the perpendicular line:	**(c)**_____
	Identify values:	**(d)** $m =$ ____; $x_1 =$ ____; $y_1 =$ _____
Step 3: Use the point-slope form of a line with the given point and the slope found in Step 2 to find the equation of the perpendicular line.	Point-slope form of a line:	**(e)**_____
	Substitute values into (e):	**(f)**_____
Step 4: Put the equation in slope-intercept form by solving for *y*.	Solve for *y*:	**(g)**_____

Do the Math Exercises 3.6
Parallel and Perpendicular Lines

In Problems 1–3, fill in the chart with the missing slopes.

	Slope of the Given Line	Slope of a Line Parallel to the Given Line	Slope of a Line Perpendicular to the Given Line
1.	$m = 4$		
2.	$m = -\dfrac{1}{8}$		
3.	$m = \dfrac{5}{2}$		

In Problems 4–7, determine if the lines are parallel, perpendicular, or neither.

4. $L_1 : y = -4x + 3$
   $L_2 : y = 4x - 1$

5. $L_1 : y = 3x - 1$
   $L_2 : y = 6 - \dfrac{x}{3}$

4. _____

5. _____

6. $L_1 : x - 4y = 24$
   $L_2 : 2x - 8y = -8$

7. $L_1 : x - 2y = -8$
   $L_2 : x + 2y = 2$

6. _____

7. _____

In Problems 8–11, find the equation of the line that contains the given point and is parallel to the given line. Write the equation in slope-intercept form, if possible.

8. $(7, -5); y = 2x + 6$

9. $(-4, 5); x = -3$

8. _____

9. _____

10. $(4, -8)$; $y = -1$

11. $(6, 7)$; $2x + 3y = 9$

10. _____

11. _____

In Problems 12–15, find the equation of the line that contains the given point and is perpendicular to the given line. Write the equation in slope-intercept form, if possible.

12. $(4, 7)$; $y = \dfrac{1}{3}x - 3$

13. $(-2, -5)$; $y = -2x + 5$

12. _____

13. _____

14. $(3, -6)$; y-axis

15. $(11, -6)$; x-axis

14. _____

15. _____

Five-Minute Warm-Up 3.7
Linear Inequalities in Two Variables

1. Solve: $x + 5 < -3$

1. _____

2. Solve: $5x - 2 \geq -17$

2. _____

3. Solve: $-3 + 7(x - 2) \geq -2(1 - 5x)$

3. _____

4. Graph: $y = 2x - 1$

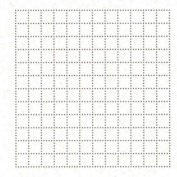

Guided Practice 3.7
Linear Inequalities in Two Variables

Objective 1: Determine Whether an Ordered Pair Is a Solution to a Linear Inequality

1. *True or False:* To determine if an ordered pair is a solution to the linear inequality, substitute the values for the variables into the inequality. If a true statement results, then the ordered pair is a solution to the inequality.

1. _____

2. Determine if $(1, 3)$ is a solution to $y \le 2x + 1$. *(See textbook Example 1)*

2. _____

Objective 2: Graph Linear Inequalities

3. To graph any linear inequality in two variables, you must first graph the corresponding equation. If the inequality is strict ($<$ or $>$), you should use a _____ line. If the inequality is nonstrict ($\le$ or $\ge$), you should use a _____ line.

4. The graph of a line separates the *xy*-plane into two _____.

5. If a test point satisfies the inequality, then every ordered pair that lies in that half plane also satisfies the

inequality. To represent this solution set, we _____ the half plane containing the test point.

6. Graph the linear inequality $y \ge x + 2$. *(See textbook Example 2)*

Step 1: We replace the inequality symbol with an equal sign and graph the corresponding line.	Write the equation:	**(a)** _____
	Identify the slope:	**(b)** _____
	Identify the *y*-intercept:	**(c)** _____
	Graph the line on the grid provided in (h). Is the line connecting these points solid or dashed?	**(d)** _____
Step 2: We select any test point that is not on the line and determine whether the test point satisfies the inequality. When the line does not contain the origin, it is usually easiest to choose the origin, (0, 0), as the test point.	Select a test point:	**(e)** _____ $y \ge x + 2$
	Substitute the values for the variables into the inequality. Is the statement true or false?	**(f)** _____
	Which half plane should be shaded; the half plane containing the test point or the opposite half plane?	**(g)** _____
	Graph the solution set by using the slope and *y*-intercept, graphing the line, and shading the appropriate half plane.	**(h)** $y \ge x + 2$

7. Graph the linear inequality $4x - y > 0$. *(See textbook Example 4)*

Replace the inequality symbol with an equal sign to obtain the graph of $4x - y > 0$. Is the line solid or dashed?

(a) _____

Select a test point. Can you use (0, 0)?

(b) _____

Determine if the inequality is satisfied by your test point? If the statement is true, shade the half plane containing the test point. If the statement is false, shade the opposite half plane.

(c)

Objective 3: Solve Problems Involving Linear Inequalities

8. Wood Flooring Mary Jo owns a discount flooring store and sells two different grades of wood flooring. There is a domestic cherry wood that sells for $4.32 per square foot and a reclaimed Douglas fir that sells for $2.50 per square foot. How many square feet of wood flooring does she need to sell to have an income of at least $5000? *(See textbook Example 6)*

(a) Write a linear inequality that describes Mary Jo's options for selling wood flooring.

Step 1: Identify We want to determine how many square feet of each she should sell to earn at least $5000. This requires an inequality in two variables.

Step 2: Name the Unknowns Let x represent the number of square feet of cherry flooring sold and let y represent the number of square feet of Douglas fir sold.

Step 3: Translate If cherry wood sells for $4.32 per square foot and Douglas fir sells for $2.50 per square foot, write an inequality the represents her total income greater than or equal to $5000.

(a) _____

(b) Will Mary Jo make her sales goal if she sells 600 square feet of cherry wood and 850 square feet of Douglas fir?

(b) _____

Sullivan/Struve/Mazzarella, *Elementary Algebra*, 2e

Do the Math Exercises 3.7
Linear Inequalities in Two Variables

In Problems 1–4, determine which of the following ordered pairs, if any, are solutions to the given linear inequality in two variables.

1. $y \leq x - 5$ $A(-4, -6)$ $B(0, -8)$ $C(3, 10)$

2. $y > -2x + 1$ $A(0, 0)$ $B(-2, 3)$ $C(2, -1)$

1. _____

2. _____

3. $3x + 5y \geq 4$ $A(2, 0)$ $B(-3, 4)$ $C(6, -2)$

4. $x > 10$ $A(3, 12)$ $B(11, -6)$ $C(10, 16)$

3. _____

4. _____

In Problems 5–12, graph each linear inequality.

5. $y \leq 4x + 2$

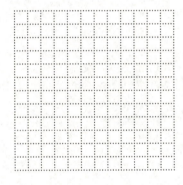

6. $y > x - 3$

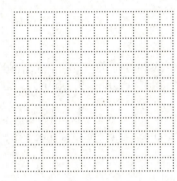

7. $y < 4$

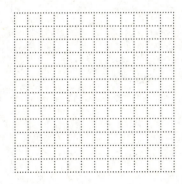

8. $y \geq \dfrac{3}{2}x - 2$

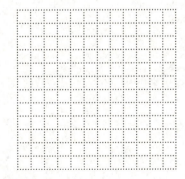

9. $2x + 6y < -6$

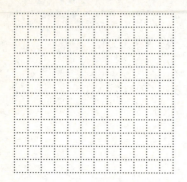

10. $6x - 8y \geq 24$

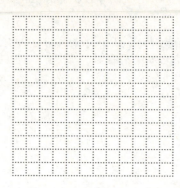

11. $x - y > 0$

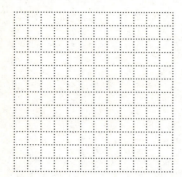

12. $x \geq 2$

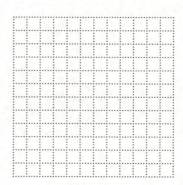

13. Fishing Trip Patrick's rowboat can hold a maximum of 500 pounds. In Patrick's circle of friends, the average adult weighs 160 pounds and the average child weighs 75 pounds.

(a) Write a linear inequality that describes the various combinations of the number of adults a and children c that can go on a fishing trip in the rowboat.

13a. _____

(b) Will the boat sink with 2 adults and 4 children?

13b. _____

(c) Will the boat sink with 1 adult and 5 children?

13c. _____

Five-Minute Warm-Up 4.1
Solving Systems of Linear Equations by Graphing

1. Graph each of the following linear equations.

    (a) $6x - 3y = -12$ **(b)** $y = -\dfrac{3}{2}x - 1$

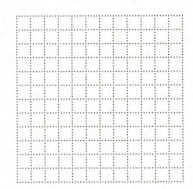   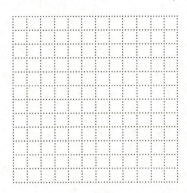

2. Determine whether $2x + 3y = -3$ is parallel to $-4x - 6y = 18$.   2.

3. Determine whether $3x + 2y = 6$ is parallel to $y = -\dfrac{3}{2}x + 3$.   3.

Guided Practice 4.1
Solving Systems of Linear Equations by Graphing

Objective 1: Determine If an Ordered Pair Is a Solution of a System of Linear Equations

1. Determine whether the given ordered pairs are solutions of the system of equations. *(See textbook Example 2)*

$$\begin{cases} x + y = 0 \\ 2x - 3y = 5 \end{cases}$$

(a) $(-2, 2)$ **(b)** $(1, -1)$ 1a. _____

       1b. _____

Objective 2: Solve a System of Linear Equations by Graphing

2. Solve the following system by graphing: $\begin{cases} x - y = -5 \\ 2x + y = -1 \end{cases}$ *(See textbook Example 3)*

Step 1: Graph the first equation in the system.	The equation $x - y = -5$ can be graphed using the intercepts or by writing the equation in slope-intercept form. We will use the intercepts. Name the x and y-intercept and then graph the line in the coordinate plane below.	**(a)** x-intercept _____ **(b)** y-intercept _____
Step 2: Graph the second equation in the system.	The equation $2x + y = -1$ can also be graphed by either method. This time we will use slope-intercept form. Write the equation in slope-intercept form. Name the slope and y-intercept and then graph the line in the coordinate plane below.	**(c)** equation _____ **(d)** slope _____ **(e)** y-intercept _____
Step 3: Determine the point of intersection of the two lines, if any.	Name the point where the lines appear to intersect.	**(f)** _____
Step 4: Verify that the point of intersection determined in Step 3 is a solution of the system.	Check that the point satisfies both of the equations in the system.	**(g)** 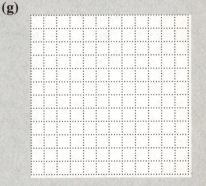

State the solution to the system: **(h)** _____

3. Solve the system by graphing: $\begin{cases} 2x + y = 1 \\ -4x - 2y = 6 \end{cases}$ *(See textbook Example 5)*

3. _____

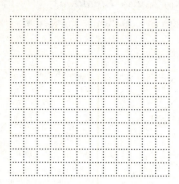

Objective 3: Classify Systems of Linear Equations as Consistent or Inconsistent

4. Complete the following chart which describes the solutions to a system of linear equations in two variables.

Number of Solutions	Classification of the System	Appearance of the Graph of the Two Lines
(a) no solution		
(b) infinitely many solutions		
(c) exactly one solution		

5. **(a)** Without graphing, determine the number of solutions of the system:

$\begin{cases} 4x - 3y = -6 \\ 8x - 6y = -12 \end{cases}$ *(See textbook Example 7)*

5a. _____

(b) State whether the system is consistent or inconsistent.

5b. _____

(c) If the system is consistent, state whether the equations are dependent or independent.

5c. _____

Do the Math Exercises 4.1
Solving Systems of Linear Equations by Graphing

In Problems 1–2, determine whether the ordered pair is a solution of the system of equations.

1. $\begin{cases} 2x+5y=0 \\ x-3y=11 \end{cases}$

   (a) $(5, -2)$
   (b) $(-5, 2)$
   (c) $(2, -3)$

2. $\begin{cases} x-4y=2 \\ 5x-8y=-6 \end{cases}$

   (a) $(-8, -2)$
   (b) $(2, 2)$
   (c) $\left(\dfrac{-10}{3}, -\dfrac{4}{3} \right)$

1a. _____

1b. _____

1c. _____

2a. _____

2b. _____

2c. _____

In Problems 3–6, solve each system of equations by graphing.

3. $\begin{cases} y=-4x+6 \\ y=-2x \end{cases}$

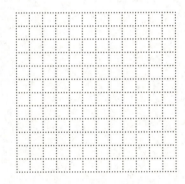

4. $\begin{cases} -2x+5y=-20 \\ 4x-10y=10 \end{cases}$

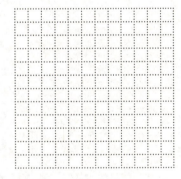

5. $\begin{cases} 2x+5y=2 \\ -x+2y=-1 \end{cases}$

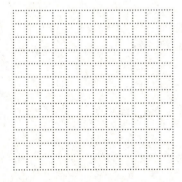

6. $\begin{cases} 3y=-4x-4 \\ 8x+2=-6y-6 \end{cases}$

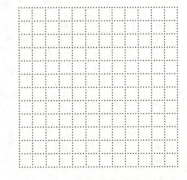

In Problems 7–10, without graphing, determine the number of solutions of each system of equations. State whether the system is consistent or inconsistent. For those systems that are consistent, state whether the equations are dependent or independent.

7. $\begin{cases} 2x + y = -3 \\ -2x - y = 6 \end{cases}$ **8.** $\begin{cases} -x + y = 5 \\ x - y = -5 \end{cases}$

7. _____

8. _____

9. $\begin{cases} y - 3 = 0 \\ x - 3 = 0 \end{cases}$ **10.** $\begin{cases} 5x + y = -1 \\ 10x + 2y = -2 \end{cases}$

9. _____

10. _____

Set up a system of linear equations in two variables that models the problem. Then solve the system of linear equations by graphing.

11. Car Rental The Speedy Car Rental agency charges $60 per day plus $0.14 per mile while the Slow-but-Cheap Car Rental agency charges $30 per day plus $0.20 per mile. Determine the number of miles for which the cost of the car rental will be the same for both companies. If you plan to drive the car for 400 miles, which company should you use?

11.  _____

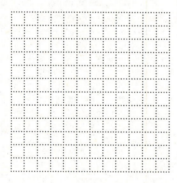

Five-Minute Warm-Up 4.2
Solving Systems of Linear Equations Using Substitution

1. Solve $5x - 3y = -18$ for y.

1. _____

2. Solve $3x + 8y = -12$ for x.

2. _____

3. Solve: $\dfrac{3}{2}x - \dfrac{5}{2}(x + 6) = -8$

3. _____

4. Solve: $-\left(\dfrac{7}{3}y + 6\right) - \dfrac{2}{3}y = -15$

4. _____

Guided Practice 4.2
Solving Systems of Linear Equations Using Substitution

Objective 1: Solve a System of Linear Equations Using the Substitution Method

1. Solve the following system by substitution: $\begin{cases} 4x + 3y = -23 & (1) \\ y = -x - 7 & (2) \end{cases}$ *(See textbook Example 1)*

Step 1: Solve one of the equations for one of the unknowns.

Equation (2) is already solve for *y*: $y = -x - 7$

Step 2: Substitute the expression for *y* into equation (1).

Equation (1): **(a)** _____

Substitute into equation (1): **(b)** _____

Step 3: Solve the equation for *x*.

Distribute the 3: **(c)** _____

Combine like terms: **(d)** _____

Add 21 to both sides: **(e)** $x =$ _____

Step 4: Substitute your value for *x* into equation 2 and solve for *y*.

Equation (2): **(f)** _____

Substitute for *x*: **(g)** _____

Solve for *y*: **(h)** _____

Step 5: Verify your solution.

Check that your ordered pair satisfies both of the equations.

State the solution: **(i)** _____

2. Solve the following system by substitution: $\begin{cases} 3x - 2y = -8 \\ x + 3y = 34 \end{cases}$ *(See textbook Example 3)*   2. _____

3. Solve the following system by substitution:
$$\begin{cases} 5x + 8y = 4 \\ y = -\dfrac{5}{8}x + \dfrac{1}{2} \end{cases}$$
(See textbook

4. _____

Example 5)

Objective 2: Solve Applied Problems Involving Systems of Linear Equations

4. Unknown Integers The sum of two numbers is -11 and their difference is -101. Find the numbers. *(See textbook Example 6)*

   Step 1: Identify This is a direct translation problem involving unknown integers.

   Step 2: Name Let the two unknown numbers be identified as x and y.

   Step 3: Translate
   (a) Write an equation showing the sum of the two numbers is -11:   4a. _____

   (b) Write an equation showing the difference of the two numbers is -101:   4b. _____

   Step 4: Solve
   (c) Write the system to be solved: 4c. _____

   (d) Solve one of the equations for one of the variables. Substitute this expression into the opposite equation and solve for the unknown. What is the solution to the system?   4d. _____

   Step 5: Check Verify that the values you found satisfy the system and make sense in the original problem.

   Step 6: Answer the question: 4e. _____

Do the Math Exercises 4.2
Solving Systems of Linear Equations Using Substitution

In Problems 1–4, solve each system of equations using substitution.

1. $\begin{cases} 5x + 2y = -5 \\ 3x - y = -14 \end{cases}$

2. $\begin{cases} y = \dfrac{2}{3}x + 1 \\ y = -\dfrac{3}{2}x + 40 \end{cases}$

1. _____

2. _____

3. $\begin{cases} y = 5x - 3 \\ y = 2x - \dfrac{21}{5} \end{cases}$

4. $\begin{cases} y = 4x \\ 2x - 3y = 5 \end{cases}$

3. _____

4. _____

In Problems 5–8, solve each system of equations using substitution. State whether the system is inconsistent, or consistent and dependent.

5. $\begin{cases} 2x - y = 3 \\ y - 2x = 3 \end{cases}$

6. $\begin{cases} y = -x + 1 \\ x + y = 1 \end{cases}$

5. _____

6. _____

7. $\begin{cases} 4 + y = 3x \\ 6x - 2y = -2 \end{cases}$

8. $\begin{cases} x + y = 2 \\ 2x + 2y = 2 \end{cases}$

7. _____

8. _____

In Problems 9–14, solve each system of equations using substitution.

9. $\begin{cases} y = \dfrac{3}{4}x \\ x = 4(y-1) \end{cases}$

10. $\begin{cases} x = 3y + 6 \\ y - \dfrac{1}{3}x = -2 \end{cases}$

9. _____

10. _____

11. $\begin{cases} 3x + 2y = 1 \\ x + 3y = -\dfrac{5}{3} \end{cases}$

12. $\begin{cases} -4x + y = 3 \\ 8x - 2y = 1 \end{cases}$

11. _____

12. _____

13. $\begin{cases} 18x - 6y = -7 \\ x + 2y = 0 \end{cases}$

14. $\begin{cases} \dfrac{2}{3}x + \dfrac{1}{3}y = 1 \\ \dfrac{1}{2}x + \dfrac{1}{4}y = \dfrac{3}{4} \end{cases}$

13. _____

14. _____

15. **Investment** Elaine wants to invest part of her $24,000 Virginia Lottery winnings in an international stock fund that yields 4% annual interest and the remainder in a domestic growth fund that yields 6.5% annually. Use the system of equations to determine the amount Elaine should invest in each account to earn $1360 interest at the end of one year, where x represents the amount invested in the international stock fund and y represents the amount invested in the domestic growth fund.

15. _____

$$\begin{cases} x + y = 24{,}000 \\ 0.04x + 0.065y = 1360 \end{cases}$$

Sullivan/Struve/Mazzarella, *Elementary Algebra*, 2e

Five-Minute Warm-Up 4.3
Solving Systems of Linear Equations Using Elimination

1. What is the additive inverse of $\dfrac{5}{4}$?   1. _____

2. What is the additive inverse of -6 ?   2. _____

3. Distribute: $-\dfrac{7}{4}\left(-12x + 8y\right)$ 3. _____

4. Solve: $-6x + 7 = 3(-2x + 1)$ 4. _____

5. Simplify: $8x - 7y + 3 + -8x - 2y - 12$ 5. _____

6. **(a)** Find the LCM of 8 and 12.   6a. _____

   (b) By what factor must 8 be multiplied so that the product of the factor and 8   6b. _____
   equal the LCM?

   (c) By what factor must 12 be multiplied so that the product of the factor and 12   6c. _____
   equal the LCM?

Guided Practice 4.3
Solving Systems of Linear Equations Using Elimination

Objective 1: Solve a System of Linear Equations Using the Elimination Method

1. This method is called "elimination" because one of the variables is eliminated through the process of addition. The elimination method is also sometimes referred to as the *addition method*.

Solve the following system by elimination: $\begin{cases} -x + y = 12 \\ x - 3y = -26 \end{cases}$ *(See textbook Example 1)*   1. _____

2. Solve the following system by elimination: $\begin{cases} 2x + y = -4 & (1) \\ 3x + 5y = 29 & (2) \end{cases}$ *(See textbook Example 2)*

Step 1: Write each equation in standard form, $Ax + By = C$.

Both equations are already in standard form.

Step 2: Our first goal is to get the coefficients on one of the variables to be additive inverses. In looking at this system, we can make the coefficients of y be additive inverses by multiplying equation (1) by -5.

Multiply both sides of (1) by -5, use the Distributive Property, and then write the equivalent system of equations.

$\begin{cases} 2x + y = -4 \\ 3x + 5y = 29 \end{cases}$

(a) $\begin{cases} \underline{\hspace{3cm}} & (1) \\ \underline{\hspace{3cm}} & (2) \end{cases}$

Step 3: We now add equations (1) and (2) to eliminate the variable y and then solve for x.

Add (1) and (2):

(b) _____

Divide both sides by -7:

(c) $x =$ _____

Step 4: Substitute your value for x into either equation (1) or equation (2). We will use equation (1) as it looks like less work.

Equation (1):

Substitute your value for x and solve for y.

$2x + y = -4$

(d) $y =$ _____

Step 5: Check your answer in both of the original equations. If both equations yield a true statement, you have the correct answer.

Write the ordered pair that is the solution to the system.

(e) _____

3. Solve the following system by elimination: $\begin{cases} \dfrac{1}{2}x - \dfrac{1}{5}y = \dfrac{2}{5} & (1) \\ -\dfrac{3}{8}x + \dfrac{5}{4}y = -\dfrac{5}{2} & (2) \end{cases}$

(See textbook Example 3)

    (a) Multiply both sides of equation (1) by the LCD of 2 and 5:   3a. _____

    (b) Multiply both sides of equation (2) by the LCD of 2, 4, and 8:   3b. _____

    (c) Solve this equivalent system by elimination and state the solution: 3c. _____

4. When solving a system of linear equations by elimination, if both variables were eliminated and the statement $0 = 6$ was left, what is the solution to this system?   4. _____
(See textbook Example 4)

5. When solving a system of linear equations by elimination, if both variables were eliminated and the statement $0 = 0$ was left, what is the solution to this system?   5. _____
(See textbook Example 5)

Objective 2: Solve Applied Problems Involving Systems of Linear Equations

6. Ticket Prices Admission to the movie theatre costs $4.50 per child and $7.50 per adult. If the theatre took in $3337.50 and had 525 patrons, how many of each ticket were sold? *(See textbook Example 6)*

    (a) If *a* represents the number of adult tickets sold and *c* represents the number of children's tickets sold, write an equation that shows that the total number of tickets sold   6a. _____
is 525.

    (b) Use the rate for a child's ticket and number of children's tickets sold plus the rate for an adult ticket and the number of adult tickets sold to write an equation showing the total   6b. _____
revenue for this performance is $3337.50.

    (c) Solve the system.   6c. _____

    (d) Answer the question in the problem.   6d. _____

Do the Math Exercises 4.3
Solving Systems of Linear Equations Using Elimination

In Problems 1–2, solve each system of equations using elimination.

1. $\begin{cases} 2x + 6y = 6 \\ -2x + y = 8 \end{cases}$ **2.** $\begin{cases} 3x + 2y = 7 \\ -3x + 4y = 2 \end{cases}$

1. _____

2. _____

In Problems 3–4, solve each system of equations using elimination. State whether the system is inconsistent or consistent and dependent.

3. $\begin{cases} x - y = -4 \\ -2x + 2y = -8 \end{cases}$ **4.** $\begin{cases} 2x + 5y = 15 \\ -6x - 15y = -45 \end{cases}$

3. _____

4. _____

In Problems 5–10, solve each system of equations using elimination.

5. $\begin{cases} 5x - y = 3 \\ -10x + 2y = 2 \end{cases}$ **6.** $\begin{cases} 2x + 3y = 2 \\ 5x + 7y = 0 \end{cases}$

5. _____

6. _____

7. $\begin{cases} x + 3y = 6 \\ 9y = -3x + 18 \end{cases}$ **8.** $\begin{cases} 5x + 7y = 6 \\ 2x - 3y = 11 \end{cases}$

7. _____

8. _____

9.
$$\begin{cases} 12x + 15y = 55 \\ \dfrac{1}{2}x + 3y = \dfrac{3}{2} \end{cases}$$

10.
$$\begin{cases} -2.4x - 0.4y = 0.32 \\ 4.2x + 0.6y = -0.54 \end{cases}$$

9. _____

10. _____

In Problems 11–14, solve by any method: graphing, substitution, or elimination.

11.
$$\begin{cases} 2x - 3y = -10 \\ -3x + y = 1 \end{cases}$$

12.
$$\begin{cases} 0.25x + 0.10y = 3.70 \\ x + y = 25 \end{cases}$$

11. _____

12. _____

13. Carbs Yvette and José go to McDonald's for breakfast. Yvette orders two sausage biscuits and one 16-ounce orange juice. The entire meal had 98 grams of carbohydrates. José orders three sausage biscuits and two 16-ounce orange juices and his meal had 168 grams of carbohydrates. How many grams of carbohydrates are in a sausage biscuit? How many grams of carbohydrates are in a 16-ounce orange juice? Solve the system

$$\begin{cases} 2b + u = 98 \\ 3b + 2u = 168 \end{cases}$$

where b represents the number of grams of carbohydrates in a sausage biscuit and u represents the number of grams of carbohydrates in an orange juice to find the answers.

13. _____

Five-Minute Warm-Up 4.4
Solving Direct Translation, Geometry, and Uniform Motion Problems Using Systems of Linear Equations

1. If you are traveling at an average speed of 65 miles per hour (mph),
    (a) how far will you travel in 5 hours?   **(b)** If you went 195 miles, how long were you traveling?

    1a. _____

    1b. _____

2. Suppose that you have $7000 in a Certificate of Deposit (CD) that pays 4.5% annual simple interest. What is the amount of interest paid
    (a) after 3 years?   **(b)** after two months?

    2a. _____

    2b. _____

3. Translate each of the following sentences into an equation. *DO NOT SOLVE.*

    (a) Seven less than twice a number is 45.   3a. _____

    (b) 5 times the sum of 3 and some number is the same as half the number.   3b. _____

    (c) The quotient of a number and 3 is equal to the number subtracted from 20.   3c. _____

Guided Practice 4.4
Solving Direct Translation, Geometry, and Uniform Motion Problems Using Systems of Linear Equations

Objective 1: Model and Solve Direct Translation Problems

1. Fun with Numbers The sum of four times a first number and a second number is 68. If the first number is decreased by twice the second number the result is -1. Find the numbers. *(See textbook Example 1)*

   Step 1: Identify We are looking for the two unknown numbers.

   Step 2: Name Let x represent the first number and y represent the second number.

   Step 3: Translate

   (a) Write the equation for the sum of four times a first number and a second number   1a. _____
   is 68:

   (b) Write the equation for the first number decreased by twice the second number   1b. _____
   results in -1:

   (c) Step 4: Solve Use either substitution or elimination to find the solution to your   1c. _____
   system of equations. What is the solution to the system?

   Step 5: Check Verify that the values you found satisfy the system and make sense in
   the original problem.

   (d) Step 6: Answer the question in the problem: 1d. _____

Objective 2: Model and Solve Geometry Problems

2. Two angles are supplementary if the sum of the measures of the angles is: 2. _____

3. If one angle is 30° more than its supplement, find the measures of the two angles. *(See textbook Example 3)*

   (a) If one angle has x degrees and the supplement has y degrees, write an equation
   showing the sum of the measures of the two angles: 3a. _____

   (b) Write an equation showing one angle is 30° more than its supplement: 3b. _____

   (c) Solve your system of two linear equations.   3c. _____

   (d) Answer the question in the problem.   3d. _____

Objective 3: Model and Solve Uniform Motion Problems

4. Airplane Speed A plane can fly 2400 miles east, with the wind, in 6 hours. The return trip west to the same point, against the wind, takes 8 hours. Find the airspeed of the plane and the effect wind resistance has on the plane. *(See textbook Example 4)*

Step 1: Identify This is a uniform motion problem. We want to determine the airspeed of the plane and the effect of wind resistance.

Step 2: Name We will let *a* represent the airspeed of the plane and *w* represent the effect of wind resistance.

Step 3: Translate
(a) Flying east, with the wind, the ground speed of the plane is: 4a. _____

(b) Flying west, against the wind, the ground speed of the plane is: 4b. _____

Complete the table from the information given, using *a* for speed of the plane and *w* for speed of the wind.

	Distance (miles)	Rate (mph)	Time (hours)
(c) With the wind			
(d) Against the wind			

(e) Write a system of equations using the information in your table: 4e. _____

(f) Step 4: Solve Use either substitution or elimination to find the solution to your system of equations.   4f. _____

Step 5: Check Does your solution satisfy each of the equations in the system? Does your answer seem reasonable?

(g) Step 6: Answer the question in the problem: 4g. _____

Do the Math Exercises 4.4
Solving Direct Translation, Geometry, and Uniform Motion Problems Using Systems of Linear Equations

Complete the system of equations. Do not solve the system.

1. The sum of two numbers is 90. If 20 is added to 3 times the smaller number, the result exceeds twice the larger number by 50. Let a represent the smaller number and let b represent the larger number.

$$\begin{cases} a + b = 90 \\ \underline{\hspace{2cm}} = \underline{\hspace{2cm}} \end{cases}$$

Complete the system of equations. Do not solve the system.

2. The perimeter of a rectangle is 212 centimeters. The length is 8 centimeters less than three times the width. Let l represent the length of the rectangle and let w represent the width of the rectangle.

$$\begin{cases} 2w + 2l = 212 \\ \underline{\hspace{2cm}} = \underline{\hspace{2cm}} \end{cases}$$

Complete the system of equations. Do not solve the system.

3. A plane flew with the wind for 3 hours, covering 1200 miles. It then returned over the same route to the airport against the wind in 4 hours. Let a represent the airspeed of the plane and w represent the effect of wind resistance.

$$\begin{cases} 3(a + w) = 1200 \\ \underline{\hspace{2cm}} = \underline{\hspace{2cm}} \end{cases}$$

4. **Fun with Numbers** Find two numbers whose sum is 55 and whose difference is 17.   4. _____

5. **Fun with Numbers** The sum of two numbers is 32. Twice the larger subtracted from   5. _____
    the smaller is –22. Find the numbers.

6. **Perimeter of a Parking Lot** A rectangular parking lot has a perimeter of 125 feet. The length of the parking lot is 10 feet more than the width. What is the length of the parking lot? What is the width?

6. _____

7. **Working with Complements** The measure of one angle is 10° less than the measure of three times its complement. Find the measures of the two angles.

7. _____

8. **Finding Supplements** The measure of one angle is 20° more than two-thirds the measure of its supplement. Find the measures of the two angles.

8. _____

9. **Southwest Airlines Plane** A Southwest Airlines plane can fly 455 mph against the wind and 515 mph when it flies with the wind. Find the effect of the wind and the groundspeed of the airplane.

9. _____

10. **Rowing** On Monday afternoon, Andrew rowed his boat with the current for 4.5 hours and covered 27 miles, stopping in the evening at a campground. On Tuesday morning he returned to his starting point against the current in 6.75 hours. Find the speed of the current and the rate at which Andrew rowed in still water.

10. _____

11. **Horseback Riding** Monica and Gabriella enjoy riding horses at a dude ranch in Colorado. They decide to go down different trails, agreeing to meet back at the ranch later in the day. Monica's horse is going 4 mph faster than the Gabriella's, and after 2.5 hours, they are 20 miles apart. Find the speed of each horse.

11. _____

Five-Minute Warm-Up 4.5
Solving Mixture Problems Using Systems of Linear Equations

1. Suppose that Sherry has a credit card balance of $2000. Each month, the credit card charges 18% annual simple interest on any outstanding balances.

   (a) What is the interest that Sherry will be charged on this loan after one month?   1a. _____

   (b) What is the balance on Sherry's credit card after one month?   1b. _____

2. Write the percent as a decimal.   2a. _____
   (a) 40%   (b) 0.15%
                                                                                               2b. _____

3. Solve by any appropriate method: $\begin{cases} 2x + 5y = -11 \\ 3x + 2y = 11 \end{cases}$ 3. _____

Guided Practice 4.5
Solving Mixture Problems Using Systems of Linear Equations

Objective 1: Draw Up a Plan for Modeling Mixture Problems

1. The cost of an adult ticket to the art museum is $7.50 and student tickets cost $4.00. When a group of 40 adults and students entered the museum, the total receipts were $202. How many adults and how many students were in the group that entered the museum? *(See textbook Example 1)*

Fill in the chart that summarizes the information in the problem.

	Number	Cost per Person in Dollars	Amount
(a) Adults			
(b) Students			
(c) Total			

(d) Write a system of equations to solve this problem. Do not solve the system.   1d. _____

Objective 2: Set Up and Solve Money Problems Using the Mixture Model

2. Johnny has $6.75 in dimes and quarters. If he has 42 coins, how many quarters does Johnny have? *(See textbook Example 2)*

Fill in the chart that summarizes the information in the problem.

	Number of Coins	Value per Coin in Dollars	Total Value
(a) Quarters			
(b) Dimes			
(c) Total			

(d) Write a system of equations to solve this problem.   2d. _____

(e) Solve the system and answer the question in the problem.   2e. _____

Objective 3: Set Up and Solve Dry Mixture and Percent Mixture Problems

3. A candy store sells chocolate-covered almonds for $6.50 per pound and chocolate-covered peanuts for $4.00 per pound. The manager decides to make a bridge mix that combines the almonds with the peanuts. She wants the bridge mix to sell for $6.00 per pound, and there should be no loss in revenue from selling the bridge mix versus the almonds and peanuts alone. How many pounds of chocolate-covered almonds and chocolate-covered peanuts are required to create 50 pounds of bridge mix? *(See textbook Example 4)*

Complete the chart that summarizes the information in the problem.

	Price $/Pound	Number of Pounds	Revenue
(a) Almonds			
(b) Peanuts			
(c) Mix			

(d) Write a system of equations to solve this problem. Do not solve the system.   3d. _____

4. The Chemistry stockroom has a jar labeled 25% hydrochloric acid and another labeled 40% hydrochloric acid. If 90 ml of 30% solution are needed for today's lab experiment, how much of each should the stockroom assistant mix together? *(See textbook Example 5)*

Complete the chart that summarizes the information in the problem.

	Number of ml	Concentration	Amount of Pure HCl
(a) 25% HCl Solution			
(b) 40% HCl Solution			
(c) 30% HCl Solution			

(d) Write a system of equations to solve this problem. Do not solve the system.   4d. _____

Name: Date:
Instructor: Section:

Do the Math Exercises 4.5
Solving Mixture Problems Using Systems of Linear Equations

In Problems 1–3, fill in the table from the information given. Then write the system that models the problem. DO NOT SOLVE.

1. John Murphy sells jewelry at art shows. He sells bracelets for $10 and necklaces for $15. At the end of one day John found that he had sold 69 pieces of jewelry and had receipts of $895.   1. _____

	Number ·	Cost per Item =	Total Value
Bracelets			
Necklaces			
Total			

2. Sherry has a savings account that earns 2.75% simple interest per year and a certificate of deposit (CD) that earns 2% simple interest annually. At the end of one year, Sherry received $37.75 in interest on a total investment of $1700 in the two accounts.   2. _____

	Principal ·	Rate =	Interest
Savings Account			
Certificate of Deposit			
Total			

3. A merchant wishes to mix peanuts worth $5 per pound and trail mix worth $2 per pound to yield 40 pounds of a nutty mixture that will sell for $3 per pound.   3. _____

	Number of Pounds ·	Price per Pound =	Total Value
Peanuts			
Trail Mix			
Total			

In Problems 4–6, complete the system of linear equations to solve the problem. Do not solve.

4. On a school field trip, 22 people attended a dress rehearsal of the Broadway play, "Avenue Q." They paid $274 for the tickets, which cost $15 for each adult and $7 for each child. How many adults and how many children attended "Avenue Q"? Let a represent the number of adult tickets purchased and let c represent the number of children's tickets purchased.

$$\begin{cases} a + c = 22 \\ \underline{} + \underline{} = 274 \end{cases}$$

Sullivan/Struve/Mazzarella, *Elementary Algebra*, 2e   161

5. You have a total of $2650 to invest. Account A pays 5% annual interest and account B pays 6.5% annual interest. How much should you invest in each account if you would like the investment to earn $155 at the end of one year? Let A represent the amount of money invested in the account that earns 5% annual interest and let B represent the amount of money invested in the account that earns 6.5% annual interest.

$$\begin{cases} A + B = \underline{\hspace{1cm}} \\ \underline{\hspace{1cm}} + \underline{\hspace{1cm}} = 155 \end{cases}$$

6. The Latte Shoppe sells Bold Breakfast coffee for $8.60 per pound and Wake-Up coffee for $5.75 per pound. One day, the amount of Bold Breakfast coffee sold was 2 pounds less than twice the amount of Wake-Up coffee and the revenue received from selling both types of coffee was $143.45. How many pounds of each type of coffee were sold that day? Let B represent the number of pounds of Bold Breakfast coffee sold and let W represent the number of pounds of Wake-Up sold. Complete the system of equations:

$$\begin{cases} 8.60B + 5.75W = \underline{\hspace{1cm}} \\ B = \underline{\hspace{1cm}} - \underline{\hspace{1cm}} \end{cases}$$

In Problems 7–9, write a system of equations.

7. **Ticket Pricing** A ticket on the roller coaster is priced differently for adults and children. One day there were 5 adults and 8 children in a group and the cost of their tickets was $48.50. Another group of 4 adults and 12 children paid $57. What is the price of each type of ticket?

7. _____

8. **Investments** Ann has $5000 to invest. She invests in two different accounts, one expected to return 4.5% and the other expected to return 9%. In order to earn $382.50 for the year, how much should she invest at each rate?

8. _____

9. **Alcohol** A laboratory assistant is asked to mix a 30% alcohol solution with 21 liters of an 80% alcohol solution to make a 60% alcohol solution. How many liters of the 30% alcohol solution should be used?

9. _____

Five-Minute Warm-Up 4.6
Systems of Linear Inequalities

1. Solve $18x + 30 \leq -6$ and write your answer in interval notation.

1. _____

2. Solve $\dfrac{5}{2}(4x - 12) < 15x - 25$ and write your answer in interval notation.

2. _____

3. Graph each of the following linear inequalities.

    (a) $y \geq -x + 2$

                                                      (b) $8x - 4y < -16$

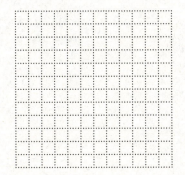

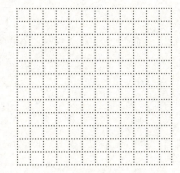

Guided Practice 4.6
Systems of Linear Inequalities

Objective 1: Determine Whether an Ordered Pair Is a Solution of a System of Linear Inequalities

1. Determine which of the following points, if any, satisfy the system of linear inequalities. *(See textbook Example 1)*

$$\begin{cases} 2x - y \geq -3 \\ x + 5y < 4 \end{cases}$$

(a) $(-1, 1)$ **(b)** $(-4, -3)$ **(c)** $(1, -1)$ 1. _____

Objective 2: Graph a System of Linear Inequalities

2. The only way to show the solution of a system of linear inequalities is by _____.

3. To graph the inequality $x - 8y > 4$, we use a _____ line as the boundry. *Refer to textbook Section 3.7.*

4. To graph the inequality $3x + 2y \geq -6$, we use a _____ line as the boundry.

5. To graph a linear inequality, we first graph the equation to determine the boundary line. This line divides the plane into two half-planes. To decide which half-plane to shade, we use a test point such as (0, 0), provided that this point does not lie on the line. If the test point satisfies the inequality, we shade the half-plane that contains the point. If the test point does not satisfy the inequality, we shade

_____.

6. When graphing a system of linear inequalities, we are looking for the ordered pairs that satisfy both

inequalties simultaneously. Therefore, the solution of the system is the _____ of the graphs of the linear inequalties.

7. Graph the system: $\begin{cases} 2x + 3y > -3 \\ -3x + y \leq 2 \end{cases}$ *(See textbook Examples 2 and 3)*

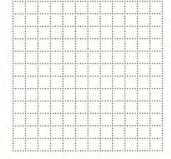

Objective 3: Solve Applied Problems Involving Systems of Linear Inequalities

8. Party Planning Alexis and Sarah are planning a barbeque for their friends. They plan to serve grilled fish and carne asada and want to spend at most $40 on the meat. The fish sells for $8 per pound and the carne asada is $5 per pound. Since most of their friends do not eat red meat, they plan to buy at least twice as much fish as carne asada. Let x represent the amount of carne asada purchased and y represent the amount of fish purchased. *(See textbook Example 5)*

(a) Write an inequality that describes how much Alexis and Sarah will spend on meat.

8a. _____

(b) Write an inequality that describes that amount of carne asada that will be purchased relative to the amount of fish that will be purchased.

8b. _____

(c) Since Alexis and Sarah will purchase a positive quantity of meat, write the two inequalities that describe this constraint.

8c. _____

_____

(d) Graph the system of inequalities.

(e) Can Alexis and Sarah purchase 5 pounds of carne asada and 12 pounds of fish and stay within their budget?

8e. _____

Do the Math Exercises 4.6
Systems of Linear Inequalities

In Problems 1–2, determine which of the following point(s), if any, is a solution of the system of linear inequalities.

1. $\begin{cases} 2x - 3y < 3 \\ 2x + y < -5 \end{cases}$

    (a) (–4, 1)

    (b) $\left(-\dfrac{3}{2}, -2\right)$

    (c) (–1, –2)

2. $\begin{cases} x - 2y > 2 \\ -3x - 2y \le 6 \end{cases}$

    (a) $\left(-1, -\dfrac{3}{2}\right)$

    (b) (0, –4)

    (c) (4, –1)

1a. _____

1b. _____

1c. _____

2a. _____

2b. _____

2c. _____

In Problems 3–6, graph each system of linear inequalities.

3. $\begin{cases} y > 2 \\ y > \dfrac{2}{3}x - 1 \end{cases}$

4. $\begin{cases} x + y < 2 \\ 3x - 5y \ge 0 \end{cases}$

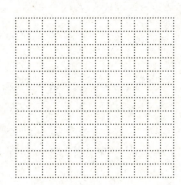

5.
$$\begin{cases} -y < \dfrac{2}{3}x + 1 \\ -3x + y \le 2 \end{cases}$$

6.
$$\begin{cases} x - y \le 3 \\ 2x + 3y \le -9 \end{cases}$$

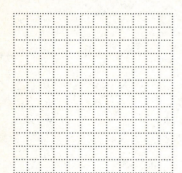

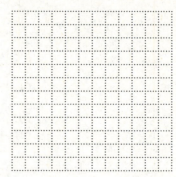

7. **Party Food** Steven and Christopher are planning a party. They plan to buy bratwurst for $4.00 per pound and hamburger patties that cost $3.00 per pound. They can spend at most $70 and think they should have no more than 20 pounds of bratwurst and hamburger patties. A system of inequalities that models the situation is given by

$$\begin{cases} 4b + 3h \le 70 \\ b + h \le 20 \\ b \ge 0 \\ h \ge 0 \end{cases}$$

where *b* represents the number of pounds of bratwurst and *h* represents the number of pounds of hamburger patties.

(a) Graph the system of linear inequalities.

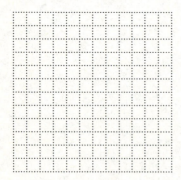

(b) Is it possible to purchase 10 pounds of bratwurst and 10 pounds of hamburger patties? 7b. _____

(c) Is it possible to purchase 15 pounds of bratwurst and 5 pounds of hamburger patties? 7c. _____

Five-Minute Warm-Up 5.1
Adding and Subtracting Polynomials

1. What is the coefficient of $-a^3$?   1. _____

2. Combine like terms: $-3x^2 + x - 5 + x^2 - x + 9$ 2. _____

3. Simplify the expression: $5 - 2(x + 3y) - 4(2x - 1)$ 3. _____

4. Use the Distributive Property to simplify: $\dfrac{5}{3}(-9x + 6)$ 4. _____

5. Simplify: $x^2 y + 6xy^2 - \left(3xy^2 + 2x^2 y\right)$ 5. _____

6. Evaluate $-\dfrac{3}{4}x^2 - 5x$ for $x = -2$.   6. _____

Guided Practice 5.1
Adding and Subtracting Polynomials

Objective 1: Define Monomial and Determine the Degree of a Monomial

1. For our study of polynomials, we begin with some definitions regarding monomials. *(See textbook Example 1)*

(a) In your own words, what is a *monomial*? _____

(b) What is the coefficient of a monomial? _____

(c) How do you determine the degree of a monomial in one variable? _____

(d) How do you determine the degree of a monomial in more than one variable? *(See textbook Example 3)*

_____

Objective 2: Define Polynomial and Determine the Degree of a Polynomial

2. Next, we move on to some definitions used with polynomials. *(See textbook Example 4)*

(a) In your own words, define *polynomial*. _____

(b) When is a polynomial in *standard form*? _____

(c) How do you determine the degree of a polynomial? _____

3. Some polynomials have special names. Always simplify the polynomial first, if possible, before determining if the polynomial has one of the following specific names:

(a) a polynomial with exactly one term is a _____

(b) a polynomial with two different monomials is a _____

(c) a polynomial with exactly three different monomials is a _____

(d) a polynomial with more than three terms is called a _____

Objective 3: Simplify Polynomials by Combining Like Terms

4. In your own words, define *like terms*. _____

5. To add two polynomials, we need to combine the like terms of the polynomials. Parentheses are included to indicate the first polynomial added to a second polynomial.

Find the sum using horizontal addition: $\left(6x^3 + 3x^2 - 5x + 2\right) + \left(x^3 - 4x^2 + 3\right)$ *(See textbook Example 6)*

5. _____

6. Find the sum using vertical addition: $\left(x^2 y - 5x^2 y^2 - 3xy^2\right) + \left(-2x^2 y - 9x^2 y^2 + 4xy^2\right)$ *(See textbook Example 7)*

6. _____

7. Find the difference: $\left(6z^3 + 2z^2 - 5\right) - \left(-3z^3 + 9z^2 - z + 1\right)$ *(See textbook Example 8)*

7. _____

8. Perform the indicated operations: $\left(6n^3 - 2n^2 + n - 1\right) - \left(4n^2 - 1\right) + \left(-6n^3 + 5n^2 + n\right)$ *(See textbook Example 10)*

8. _____

Objective 4: Evaluate Polynomials

9. Evaluate the polynomial $2x^3 - x^2 - 3x + 5$ for *(See textbook Example 11)*

   (a) $x = -1$ **(b)** $x = 3$ 9a. _____

9b. _____

10. Evaluate the polynomial $-a^2 b^2 + 2ab - b^3$ for $a = 2$ and $b = -1$. *(See textbook Example 12)*   10. _____

Do the Math Exercises 5.1
Adding and Subtracting Polynomials

In Problems 1–2, determine whether the given expression is a monomial (Yes or No). For those that are monomials, state the coefficient and degree.

1. $-x^6 y$

2. y^{-1}

1. _____

2. _____

In Problems 3–4, determine whether the algebraic expression is a polynomial (Yes or No). If it is a polynomial, write the polynomial in standard form, determine its degree, and state if it is a monomial, binomial, or trinomial. If it is a polynomial with more than three terms, identify the expression as a polynomial.

3. $4y^{-2} + 6y - 1$

4. $p^5 - 3p^4 + 7p + 8$

3. _____

4. _____

In Problems 5–8, add the polynomials. Express your answer in standard form.

5. $(x^2 - 2) + (6x^2 - x - 1)$

6. $(3 - 12w^2) + (2w^2 - 5 + 6w)$

5. _____

6. _____

7. $\left(\dfrac{3}{8}b^2 - \dfrac{3}{5}b + 1\right) + \left(\dfrac{5}{6}b^2 + \dfrac{2}{15}b - 1\right)$

8. $(4a^2 + ab - 9b^2) + (-6a^2 - 4ab + b^2)$

7. _____

8. _____

In Problems 9–12, subtract the polynomials. Express your answer in standard form.

9. $(3x^2 + x - 3) - (x^2 - 2x + 4)$

10. $(m^4 - 3m^2 + 5) - (3m^4 - 5m^2 - 2)$

9. _____

10. _____

11. $\left(\dfrac{7}{4}x^2 - \dfrac{5}{8}x - 1\right) - \left(\dfrac{7}{6}x^2 + \dfrac{5}{12}x + 5\right)$

12. $(4m^2n - 2mn - 4) - (10m^2n - 6mn - 3)$

11. _____

12. _____

In Problems 13–16, evaluate the polynomials for each of the given value(s).

13. $-x^2 + 10$

    (a) $x = 0$

    (b) $x = -1$

    (c) $x = 1$

14. $2 + \dfrac{1}{2}n^2$

    (a) $n = 4$

    (b) $n = 0.5$

    (c) $n = -\dfrac{1}{4}$

13a. _____

13b. _____

13c. _____

14a. _____

14b. _____

14c. _____

15. $-2ab^2 - 2a^2b - b^3$

    for $a = 1$ and $b = -2$

16. $m^2n^2 - mn^2 + 3m^2 - 2$

    for $m = \dfrac{1}{2}$ and $n = -1$

15. _____

16. _____

Sullivan/Struve/Mazzarella, *Elementary Algebra*, 2e

Five-Minute Warm-Up 5.2
Multiplying Monomials: The Product and Power Rules

1. Write each expression in exponential form.

(a) $\left(\dfrac{3}{4}\right) \bullet \left(\dfrac{3}{4}\right) \bullet \left(\dfrac{3}{4}\right)$

(b) $(-5)(-5)(-5)(-5)$

1a. _____

1b. _____

2. Evaluate each exponential expression.

(a) -3^2

(b) $\left(\dfrac{3}{2}\right)^3$

(c) $(-4)^2$

2a. _____

2b. _____

2c. _____

(d) $\left(-\dfrac{4}{3}\right)^3$

(e) $(-2)^5$

(f) $-(-6)^2$

2d. _____

2e. _____

2f. _____

3. Evaluate: $-\dfrac{5}{3} \bullet (-18) \bullet \left(-\dfrac{2}{5}\right)$

3. _____

4. Use the Distributive Property to simplify: $5(4 - 2x)$

4. _____

Guided Practice 5.2
Multiplying Monomials: The Product and Power Rules

Objective 1: Simplify Exponential Expressions Using the Product Rule

1. To multiply exponential expressions which have the same base, the product will have the common base and an exponent which is the _____ of the exponents of the two factors.

1. _____

2. Simplify each expression. *(See textbook Examples 2 and 3)*

    (a) $r^3 \bullet r^5$ **(b)** $m^2 \bullet n \bullet n^4 \bullet m$

2a. _____

2b. _____

Objective 2: Simplify Exponential Expressions Using the Power Rule

3. To raise exponential expressions containing a power to a power, keep the base and _____ the powers.

3. _____

4. Simplify each expression. Write the answer in exponential form. *(See textbook Examples 4 and 5)*

    (a) $\left(p^6\right)^3$ **(b)** $\left[(-x)^3\right]^4$

4a. _____

4b. _____

5. Raising a negative number to an even exponent results in a _____ number.

5. _____

6. Raising a negative number to an odd exponent results in a _____ number.

6. _____

Objective 3: Simplify Exponential Expressions Containing Products

7. When we raise a product to a power, each factor is _____.

8. Simplify each expression. *(See textbook Example 6)*

    (a) $\left(5xy^2\right)^3$ **(b)** $\left(-m^2n^3\right)^4$

8a. _____

8b. _____

Objective 4: Multiply a Monomial by a Monomial

9. Multiply and simplify each expression. *(See textbook Example 7)*

(a) $\left(-x^4\right)\left(12x^7\right)$

(b) $\left(\dfrac{7}{4}q^2\right)\left(\dfrac{12}{21}q^5\right)$

9a. _____

9b. _____

10. Multiply and simplify each expression. *(See textbook Example 8)*

(a) $\left(-5m^4n^4\right)\left(-2mn^3\right)$

(b) $\left(8pq^4\right)\left(-2p^2\right)\left(-3pq^6\right)$

10a. _____

10b. _____

Do the Math Exercises 5.2
Multiplying Monomials: The Product and Power Rules

In Problems 1–4, simplify each expression.

1. $3 \bullet 3^3$

2. $a^3 \bullet a^7$

1. _____

2. _____

3. $b^5 \bullet b \bullet b^7$

4. $(-z)(-z)^5$

3. _____

4. _____

In Problems 5–8, simplify each expression.

5. $(3^2)^2$

6. $[(-n)^7]^3$

5. _____

6. _____

7. $(k^8)^3$

8. $[(-a)^6]^3$

7. _____

8. _____

In Problems 9–12, simplify each expression.

9. $(4y^3)^2$

10. $\left(\dfrac{3}{4}n^2\right)^3$

9. _____

10. _____

11. $(-4ab^2)^3$

12. $(-3a^6bc^4)^4$

11. _____

12. _____

In Problems 13–16, multiply the monomials.

13. $(7b^5)(-2b^4)$

14. $\left(\dfrac{3}{8}y^5\right)\left(\dfrac{4}{9}y^6\right)$

13. _____

14. _____

15. $(6a^2b^3)(2a^5b)$

16. $\left(\dfrac{1}{2}b\right)(-20a^2b)\left(-\dfrac{2}{3}a\right)$

15. _____

16. _____

Five-Minute Warm-Up 5.3
Multiplying Polynomials

In Problems 1 – 6, simplify each expression.

1. $x^6 \cdot x$

2. $\left(5y^2\right)\left(4y^6\right)$

3. $\left(2z\right)^6$

4. $\left(-3a^4\right)^2$

5. $\left(\dfrac{5}{2}x\right)^2$

6. $\left(-4xy^2\right)^2\left(-x^3y^4\right)^3$

7. Use the Distributive Property to simplify: $11\left(4a - 3b\right)$

8. Simplify: $\dfrac{2}{3}\left(\dfrac{6x}{5} + \dfrac{9}{8}\right)$

1. _____

2. _____

3. _____

4. _____

5. _____

6. _____

7. _____

8. _____

Guided Practice 5.3
Multiplying Polynomials

Objective 1: Multiply a Polynomial by a Monomial
1. To multiply a monomial and a polynomial we use _____.

2. Multiply and simplify. *(See textbook Examples 2 and 3)*

(a) $-3ab^2\left(5a^2 - 2ab + 3b^2\right)$

(b) $\left(\dfrac{7}{3}x^2 - \dfrac{1}{6}x - \dfrac{4}{9}\right)18x^5$

2a. _____

2b. _____

Objective 2: Multiply Two Binomials Using the Distributive Property
3. Find the product using the Distributive Property. *(See textbook Examples 4 and 5)*

(a) $(x-4)(x-2)$

(b) $(3x-5)(2x+1)$

3a. _____

3b. _____

Objective 3: Multiply Two Binomials Using the FOIL Method

4. (a) What process does FOIL tell how to do? _____

(b) What does each of the letters stand for?

F _____ O _____ I _____ L _____

(c) Can the FOIL method be extended to multiply other types of polynomials? _____

5. Use the FOIL Method to find the product. *(See textbook Examples 6 – 9)*

(a) $(u-5)(u+3)$

(b) $(7x-2y)(x-3y)$

5a. _____

5b. _____

Objective 4: Multiply the Sum and Difference of Two Terms
6. Find the product: $(A-B)(A+B) =$

6. _____

7. Find each product. *(See textbook Example 11)*

    (a) $(x-9)(x+9)$ **(b)** $(2a+5b)(2a-5b)$ 7a. _____

                                                                     7b. _____

Objective 5: Square a Binomial

8. Find the product:

    (a) $(A+B)^2 =$ _____   **(b)** $(A-B)^2 =$ _____

9. Find the product: $(x-3)^2$ *(See textbook Example 12)*

Step 1: Use the $(A-B)^2$ pattern.

                             Write out the pattern for
                             $(A-B)^2$:   **(a)** _____

                             Identify the components. What is
                             the expression for A?   **(b)** $A =$ _____

                             What is the expression for B?   **(c)** $B =$ _____

Step 2: Find each product.   Find each of the components in   **(d)** $A^2 =$ _____

                             the pattern.

                                                             (e) $2AB =$ _____

                                                             (f) $B^2 =$ _____

                             Use the pattern to write the
                             product: $(x-3)^2 =$

                                                                 (g) _____

10. Find the product using the special products patterns. *(See textbook Examples 13 and 14)*

    (a) $(12-x)^2$ **(b)** $(4x+5y)^2$ 10a. _____

                                                                10b. _____

Objective 6: Multiply a Polynomial by a Polynomial

11. When we find the product of two polynomials, we make repeated use of the Extended Form of the Distributive Property. These multiplication problems can be formatted either horizontally, like textbook Example 15, or vertically, like textbook Example 16. Try each method and see which you prefer.

Find the product. *(See textbook Examples 15 and 16)*

    (a) $(3x-1)(x^2+2x-4)$ **(b)** $(4x+3)(2x^2-5x-3)$ 11a. _____

                                                                 11b. _____

Do the Math Exercises 5.3
Multiplying Polynomials

In Problems 1–4, use the Distributive Property to find each product.

1. $3m(2m-7)$

2. $\dfrac{3}{5}b(15b-5)$

1. _____

2. _____

3. $4w(2w^2+3w-5)$

4. $(7r+3s)(2r^2s)$

3. _____

4. _____

In Problems 5–6, use the Distributive Property to find each product.

5. $(n-7)(n+4)$

6. $(5n-2y)(2n-3y)$

5. _____

6. _____

In Problems 7–10, find the product using the FOIL method.

7. $(x+3)(x+7)$

8. $(3z-2)(4z+1)$

7. _____

8. _____

9. $(x^2-5)(x^2-2)$

10. $(3r+5s)(6r+7s)$

9. _____

10. _____

In Problems 11–12, find the product of the sum and difference of two terms.

11. $(6r-1)(6r+1)$ **12.** $(8a-5b)(8a+5b)$ 11. _____

12. _____

In Problems 13–16, find the product.

13. $(x+4)^2$ **14.** $(6b-5)^2$ 13. _____

14. _____

15. $(3x+2y)^2$ **16.** $\left(y-\dfrac{1}{3}\right)^2$ 15. _____

16. _____

In Problems 17–20, find the product.

17. $(3a-1)(2a^2-5a-3)$ **18.** $-\dfrac{4}{3}k(k+7)(3k-9)$ 17. _____

18. _____

19. $(2m^2-m+4)(-m^3-2m-1)$ **20.** $(2a-1)(a+4)(a+1)$ 19. _____

20. _____

Five-Minute Warm-Up 5.4
Dividing Monomials: The Quotient Rule and Integer Exponents

1. Find the product: $\left(7x^3 y\right)^2$

1. _____

2. Evaluate: $\left(\dfrac{6}{5}\right)^2$

2. _____

3. Find the reciprocal:

(a) -3 (b) $\dfrac{7}{8}$ 3a. _____

3b. _____

4. Simplify: $\dfrac{72}{40}$

4. _____

5. Simplify: $\dfrac{3}{8} - \dfrac{5}{6}$

5. _____

Guided Practice 5.4
Dividing Monomials: The Quotient Rule and Integer Exponents

Objective 1: Simplify Exponential Expressions Using the Quotient Rule

1. To divide exponential expressions which have the same base, the quotient will have the common base and to determine the exponent, _____ the exponent in the denominator from the exponent in the numerator.

1. _____

2. Simplify each expression. *(See textbook Examples 1 and 2)*

(a) $\dfrac{x^{11}}{x^4}$

(b) $\dfrac{-27a^5b^2}{12a^3b}$

2a. _____

2b. _____

Objective 2: Simplify Exponential Expressions Using the Quotient to a Power Rule

3. In your own words, describe the rule for raising a quotient to a power. State (a) if you should simplify the quotient first, if possible, and then raise this quotient to a power; (b) apply the power first, and then simplify, if possible; or (c) it does not matter which occurs first. Be sure to state any restrictions on your rule.

4. Simplify each expression. *(See textbook Examples 3 and 4)*

(a) $\left(-\dfrac{v}{2}\right)^3$

(b) $\left(\dfrac{28a^6}{16ab^4}\right)^2$

4a. _____

4b. _____

Objective 3: Simplify Exponential Expressions Using Zero as an Exponent

5. If a is a nonzero real number, $a^0 =$ _____.

5. _____

6. Simplify each expression. *(See textbook Example 5)*

(a) n^0

(b) $\left(9p^2\right)^0$

(c) $12x^0$

6a. _____

6b. _____

6c. _____

(d) -8^0

(e) $(-6x)^0$

(f) $-4x^0$ if $x = 2$

6d. _____

6e. _____

6f. _____

Objective 4: *Simplify Exponential Expressions Involving Negative Exponents*

7. If n is a positive integer and if a is a nonzero real number, then

(a) $a^{-n} =$ _____   (b) $\dfrac{1}{a^{-n}} =$ _____   (c) $\left(\dfrac{a}{b}\right)^{-n} =$ _____   7a. _____

7b. _____

7c. _____

From this point forward, assume all variables are nonzero. Always leave your final answer with only positive exponents.

8. Simplify each expression. *(See textbook Examples 6 – 8)*

(a) $(-2)^{-5}$ (b) $6^{-2} - 3^{-2}$ (c) $5a^{-2}$ 8a. _____

8b. _____

8c. _____

Objective 5: *Simplify Exponential Expressions Using the Laws of Exponents*

9. Simplify: $\left(\dfrac{4}{3}x^{-2}y^{4}\right)\left(-18x^{5}y^{-7}\right)$ *(See textbook Example 13)*

Step 1: Rearrange factors.

Use the commutative and associative properties of multiplication to group together the coefficients, the variables with base x and the variables with base y:

$\left(\dfrac{4}{3}x^{-2}y^{4}\right)\left(-18x^{5}y^{-7}\right) =$

(a) _____

Step 2: Find each product.

Evaluate the product of the coefficients:

(b) _____

Use $a^{n} \bullet a^{m} = a^{n+m}$ for x:

(c) _____

Use $a^{n} \bullet a^{m} = a^{n+m}$ for y:

(d) _____

Step 3: Simplify. Write the product so that the exponents are positive.

Use $a^{-n} = \dfrac{1}{a^{n}}$ for y:

(e) _____

Do the Math Exercises 5.4
Dividing Monomials: The Quotient Rule and Integer Exponents

In Problems 1–4, use the Quotient Rule to simplify. All variables are nonzero.

1. $\dfrac{10^5}{10^2}$

2. $\dfrac{x^{20}}{x^{14}}$

3. $\dfrac{-36x^2y^5}{24xy^4}$

4. $\dfrac{-15r^9s^2}{-5r^8s}$

1. _____

2. _____

3. _____

4. _____

In Problems 5–8, use the Quotient to a Power Rule to simplify. All variables are nonzero.

5. $\left(\dfrac{4}{9}\right)^2$

6. $\left(\dfrac{7}{x^2}\right)^2$

7. $\left(-\dfrac{a^3}{b^{10}}\right)^5$

8. $\left(\dfrac{2mn^2}{q^3}\right)^4$

5. _____

6. _____

7. _____

8. _____

In Problems 9–12, use the Zero Exponent Rule to simplify. All variables are nonzero.

9. -100^0

10. $\left(\dfrac{2}{5}\right)^0$

9. _____

10. _____

11. $(-11xy)^0$

12. $-11xy^0$

11. _____

12. _____

In Problems 13–16, use the Negative Exponent Rules to simplify. Write answers with only positive exponents. All variables are nonzero.

13. $-7z^{-5}$

14. $\left(\dfrac{4}{p^2}\right)^{-2}$

13. _____

14. _____

15. $\dfrac{4}{b^{-3}}$

16. $\dfrac{9}{(4t)^{-1}}$

15. _____

16. _____

In Problems 17–20, use the Laws of Exponents to simplify. Write answers with positive exponents. All variables are nonzero.

17. $\left(\dfrac{3}{5}ab^{-3}\right)\left(\dfrac{5}{3}a^{-1}b^3\right)$

18. $\dfrac{30ab^{-4}}{15a^{-1}b^{-2}}$

17. _____

18. _____

19. $(5x^{-4}y^3)^{-2}$

20. $(5a^3b^{-4})\left(\dfrac{2ab^{-1}}{b^{-3}}\right)^2$

19. _____

20. _____

Five-Minute Warm-Up 5.5
Dividing Polynomials

1. Find the quotient: $\dfrac{-135x^5}{5x}$

1. _____

2. Find the quotient: $\dfrac{12n}{108n^4}$

2. _____

3. Find the product: $-9z\left(3z^2 + 2z - 1\right)$

3. _____

4. Given the polynomial $5x + x^4 - 3x^2 - 7x^3 - x^4 + 13$,

    (a) state the degree.

4a. _____

    (b) Write the polynomial in standard form.

4b. _____

5. Simplify: $\dfrac{3}{4} + \dfrac{2}{4} - \dfrac{9}{4} = \dfrac{3 + 2 - 9}{4} =$

5. _____

6. Subtract: $x^2 + 5x - 2 - \left(x^2 + 6x\right)$

6. _____

Guided Practice 5.5
Dividing Polynomials

Objective 1: Divide a Polynomial by a Monomial

To divide a polynomial by a monomial, divide each of the terms of the polynomial numerator (dividend) by the monomial denominator (divisor). We will be using the Quotient Rule for Exponents and the fact that $\frac{a+b}{c} = \frac{a}{c} + \frac{b}{c}$. **You should continue to reduce fractions to lowest terms, use only positive exponents, and assume all variables are nonzero.**

1. Divide and simplify each expression. *(See textbook Examples 1 – 3)*

(a) $\dfrac{8x^4 + 36x^3}{4x^2}$

(b) $\dfrac{5x^2y + 20xy^3 - 5xy}{5xy}$

1a. _____

1b. _____

Objective 2: Divide a Polynomial by a Binomial

2. How do you check that long division has been done correctly?

To divide the polynomials, we use a process called long division. This is exactly the same process that you use to divide integers. Due to the tremendous popularity of hand-held calculators, this particular skill has diminished over the years. It is important to practice this process so that the steps are easy to recall while performing long division on polynomials.

3. Divide 8047 by 71. *(See textbook Example 4)*

(a) Name the divisor.

3a. _____

(b) Name the dividend.

3b. _____

(c) Perform the division. What is the quotient?

3c. _____

(d) What is the remainder?

3d. _____

(e) Express the quotient as a mixed number.

3e. _____

4. Find the quotient and remainder when $\left(x^3 - 2x^2 + x + 6\right)$ is divided by $(x+1)$. *(See textbook Example 5)*

Step 1: Divide the leading term of the dividend, x^3, by the leading term of the divisor, x. Enter the result over the term x^3.

(a) $\dfrac{x^3}{x} = $ _____

Step 2: Multiply ☐ by $x + 1$. Be sure to vertically align like terms.

(b) $x+1\overline{\smash{)}x^3 - 2x^2 + x + 6}$

Step 3: Subtract your product from the dividend.

(c) $x+1\overline{\smash{)}x^3 - 2x^2 + x + 6}$ with x^2 over the bar

Subtract and continue with Steps 4 and 5:

Step 4: Repeat Steps 1 – 3 treating $-3x^2 + x + 6$ as the dividend and dividing x into $-3x^2$ to obtain the next term in the quotient.

Step 5: Repeat Steps 1 – 3 treating $4x + 6$ as the dividend and dividing x into $4x$ to obtain the next term in the quotient.

When the degree of the remainder is less than the degree of the divisor, you are finished dividing.

Express the answer as the

$\text{Quotient} + \dfrac{\text{Remainder}}{\text{Divisor}}$.

(d) _____

Step 6: Check Verify that (Quotient)(Divisor) + Remainder = Dividend.

We leave it to you to verify the solution.

5. Before beginning long division, be sure that the divisor and dividend are written in _____ form.

5. _____

6. *True or False:* The following alternate approach correctly finds the quotient when a polynomial is divided by a binomial. $\dfrac{2x^4 - 8x}{2x + 4} = \dfrac{\cancel{2}x^4 - \cancel{8}x}{\cancel{2}x + \cancel{4}} = x^3 - 2x$

6. _____

Do the Math Exercises 5.5
Dividing Polynomials

In Problems 1–8, divide and simplify.

1. $\dfrac{3x^3 - 6x^2}{3x^2}$

2. $\dfrac{16m^3 + 8m^2 - 4}{8m^2}$

1. _____

2. _____

3. $\dfrac{16a^5 - 12a^4 + 8a^3}{20a^4}$

4. $\dfrac{7p^4 + 21p^3 - 3p^2}{-6p^4}$

3. _____

4. _____

5. $\dfrac{-5y^2 + 15y^4 - 16y^5}{5y^2}$

6. $\dfrac{35x + 20y}{-5}$

5. _____

6. _____

7. $\dfrac{21y^2 + 35x^2}{-7x^2}$

8. $\dfrac{16m^2n^3 - 24m^4n^3}{-8m^3n^4}$

7. _____

8. _____

In Problems 9–14, find the quotient using long division.

9. $\dfrac{x^2 + 4x - 32}{x - 4}$

10. $\dfrac{x^3 - x^2 - 40x + 12}{x + 6}$

9. _____

10. _____

11. $\dfrac{x^4 - 2x^3 + x^2 + x - 1}{x - 1}$

12. $\dfrac{x^3 - 7x^2 + 15x - 11}{x - 3}$

11. _____

12. _____

13. $\dfrac{9x^2 - 14}{2 + 3x}$

14. $\dfrac{64x^6 - 27}{4x^2 - 3}$

13. _____

14. _____

Five-Minute Warm-Up 5.6
Applying Exponent Rules: Scientific Notation

1. Find the product: $(2x^3)(9.5x^2)$

1. _____

2. Find the product: $(1.6n^{-2})(1.6n^{-4})$

2. _____

3. Find the quotient: $\dfrac{0.04p^{-3}}{0.2p^{-7}}$

3. _____

4. Multiply each of the following:

    (a) 4.2×1000 **(b)** 3.05×0.001 **(c)** $7 \times 3 \times 0.1$

4a. _____

4b. _____

4c. _____

5. Simplify each of the following:

    (a) $\dfrac{10^6}{10^{-2}}$ **(b)** $\dfrac{10^{-5}}{10^4}$ **(c)** $\dfrac{10^{-1}}{10^{-6}}$

5a. _____

5b. _____

5c. _____

    (d) $10^{-5} \times 10^{-3}$ **(e)** $10^{-4} \times 10^9$

5d. _____

5e. _____

Guided Practice 5.6
Applying Exponent Rules: Scientific Notation

Objective 1: Convert Decimal Notation to Scientific Notation

1. A number is written in scientific notation when it is of the form $x \times 10^N$.

   (a) x must be _____   1a. _____

   (b) N is _____   1b. _____

   (c) When 3,700 is written in scientific notation, the power of 10 is _____ (positive or   1c. _____
   negative).
                                                                  1d. _____
   (d) When 0.002 is written in scientific notation, the power of 10 is _____ (positive or
   negative).

2. Write 45,000,000 in scientific notation. *(See textbook Example 1)*

   Step 1: The "understood" decimal point in 45,000,000 follows the last 0.
   Therefore, we will move the decimal point to the left until it is between
   the 4 and the 5. Do you see why? This requires that we move the   (a) _____
   decimal $N =$ _____ places.

   Step 2: The original number is greater than 1, so the power of 10 is
   _____ (positive or negative).   (b) _____

   Now write 45,000,000 in scientific notation.   (c) _____

3. Write 0.00003 in scientific notation. *(See textbook Example 2)*

   Step 1: Because 0.00003 is less than 1, we shall move the decimal
   point until it is to the right of the 3. This requires that we move the
   decimal $N =$ _____ places.   (a) _____

   Step 2: The original number is between 0 and 1, so the power of 10 is
   _____ (positive or negative).   (b) _____

   Now write 0.00003 in scientific notation.   (c) _____

Objective 2: Convert Scientific Notation to Decimal Notation

4. To convert a number from scientific notation to decimal notation, determine the
   exponent, N, on the number 10.

   (a) If the exponent is positive, move the decimal N decimal places to the _____.   4a. _____

   (b) If the exponent is negative, then move the decimal $|N|$ decimal places to the _____.   4b. _____

5. Write 6.02×10^4 in decimal notation. *(See textbook Example 3)*

Step 1: Determine the exponent on the number 10.

(a) _____

Step 2: Since the exponent is positive, we move the decimal point *N* places to the _____.

(b) _____

Now write 6.02×10^4 in decimal notation.

(c) _____

6. Write 9.1×10^{-3} in decimal notation. *(See textbook Example 4)*

Step 1: Determine the exponent on the number 10.

(a) _____

Step 2: Since the exponent is negative, we move the decimal point $|N|$ places to the _____.

(b) _____

Now write 9.1×10^{-3} in decimal notation.

(c) _____

Objective 3: Use Scientific Notation to Multiply and Divide

7. Perform the indicated operation. Express the answer in scientific notation. *(See textbook Examples 5 and 6)*

(a) $\left(2 \times 10^5\right) \cdot \left(4 \times 10^8\right)$

(b) $\left(9 \times 10^{-15}\right) \cdot \left(6 \times 10^{10}\right)$

7a. _____

7b. _____

8. Perform the indicated operation. Express the answer in scientific notation. *(See textbook Example 7)*

(a) $\dfrac{\left(8 \times 10^{12}\right)}{\left(2 \times 10^7\right)}$

(b) $\dfrac{\left(7.2 \times 10^{24}\right)}{\left(8 \times 10^{-13}\right)}$

8a. _____

8b. _____

Do the Math Exercises 5.6
Applying Exponent Rules: Scientific Notation

In Problems 1–6, write each number in scientific notation.

1. 8,000,000,000   **2.** 0.0000001

1. _____

2. _____

3. 0.0000283   **4.** 401,000,000

3. _____

4. _____

5. 8   **6.** 120

5. _____

6. _____

In Problems 7–12, write each number in decimal notation.

7. 3.75×10^2 **8.** 6×10^6

7. _____

8. _____

9. 5×10^{-4} **10.** 4.9×10^{-1}

9. _____

10. _____

11. 5.4×10^5 **12.** 5.123×10^{-3} 11. _____

12. _____

In Problems 13–18, perform the indicated operations. Express your answer in scientific notation.

13. $(3 \times 10^{-4})(8 \times 10^{-5})$ **14.** $(4 \times 10^7)(2.5 \times 10^{-4})$ 13. _____

14. _____

15. $\dfrac{6 \times 10^3}{1.2 \times 10^5}$ **16.** $\dfrac{4.8 \times 10^7}{1.2 \times 10^2}$ 15. _____

16. _____

17. $\dfrac{0.000275}{2500}$ **18.** $\dfrac{24,000,000,000}{0.00006 \times 2000}$ 17. _____

18. _____

Five-Minute Warm-Up 6.1
Greatest Common Factor and Factoring by Grouping

1. Write as the product of prime numbers.
   (a) 36   **(b)** 225   1a. _____

   1b. _____

2. Use the Distributive Property to simplify: $-3(4x - 9)$ 2. _____

3. Find the product: $(x + 4)(2x - 3)$ 3. _____

4. Find the product: $4x^2(x - 3)(x + 3)$ 4. _____

5. Given $\dfrac{1}{2} \bullet 10 = 5$,

   (a) list the factors.   5a. _____

   (b) identify the product.   5b. _____

6. Identify the missing factor: $10x^4 \bullet \; ? = 40x^6 y$ 6. _____

Guided Practice 6.1
Greatest Common Factor and Factoring by Grouping

Objective 1: Find the Greatest Common Factor of Two or More Expressions

1. The *greatest common factor (GCF)* of a list of polynomials is the largest expression that divides evenly into all of the polynomials.

Find the GCF of 27, 36 and 45. *(See textbook Examples 1 and 2)*

Step 1: Write each number as a product of prime factors.

(a) $27 =$ _____

(b) $36 =$ _____

(c) $45 =$ _____

Step 2: Determine the common prime factors.

(d) common factors are: _____

Step 3: Find the product of the common factors found in Step 2. This number is the GCF.

(e) GCF = _____

2. Find the greatest common factor (GCF) of each of the following. *(See textbook Examples 3 and 4)*

(a) $12x^2, 30x^6$

(b) $6a^2b^3, 15ab^4c, 27a^3bc^2$

2a. _____

2b. _____

Objective 2: Factor Out the Greatest Common Factor in Polynomials

3. Factor out the great common factor: $6m^4n^2 + 18m^3n^4 - 22m^2n^5$ *(See textbook Example 6)*

Step 1: Find the GCF.

(a) GCF = _____

Step 2: Rewrite each term as the product of the GCF and remaining factor.

(b) $6m^4n^2 + 18m^3n^4 - 22m^2n^5 =$ _____

Step 3: Factor out the GCF.

(c) $6m^4n^2 + 18m^3n^4 - 22m^2n^5 =$ _____

Step 4: Check   Distribute to verify that the factorization is correct.

4. Factor out the greatest common factor. *(See textbook Examples 9 and 10)*

(a) $-24x^5 - 9x^3$

(b) $4x(x+3) + 7(x+3)$

4a. _____

4b. _____

Objective 3: *Factor Polynomials by Grouping*

5. Factor by grouping is commonly used when a polynomial has _____ terms.

5. _____

6. *True or False:* You may need to rearrange the terms in order to be able to identity a common binomial factor.

6. _____

7. Factor by grouping: $2x^2 - 4x + 3xy - 6y$ *(See textbook Example 11)*

Step 1: Group terms with common factors. In this problem the first two terms have a common factor and the last two terms have a common factor.

    (a) common factor of $2x^2 - 4x$: _____

    (b) common factor of $3xy - 6y$: _____

Step 2: In each grouping, factor out the common factor.

    (c) $2x^2 - 4x + 3xy - 6y =$ _____

Step 3: Factor out the common factor that remains.

    (d) $2x^2 - 4x + 3xy - 6y =$ _____

Step 4: Check   Multiply to verify that the factorization is correct.

8. Factor by grouping: $9x - 18y - 4ax + 8ay$ *(See textbook Example 12)*

8. _____

9. Factor by grouping: $4x^3 + 6x^2 - 12x - 18$ *(See textbook Example 13)*

9. _____

Do the Math Exercises 6.1
Greatest Common Factor and Factoring by Grouping

In Problems 1–5, find the greatest common factor, GCF, of each group of expressions.

1. 35, 42, 63

2. $26xy^2, 39x^2y$

1. _____

2. _____

3. $2x^2yz, xyz^2, 5x^3yz^2$

4. $8(x+y)$ and $9(x+y)$

3. _____

4. _____

5. $6(a-b)$ and $15(a-b)^3$

5. _____

In Problems 6–11, factor the GCF from the polynomial.

6. $-9b^2 - 6ab$

7. $8a^3b^2 + 12a^5b^2$

6. _____

7. _____

8. $5x^4 + 10x^3 - 25x^2$ **9.** $-2y^2 + 10y - 14$ 8. _____

                                                                                  9. _____

10. $-22n^4 + 18n^2 + 14n$ **11.** $a(a-5) + 6(a-5)$ 10. _____

                                                                                  11. _____

In Problems 12–15, factor by grouping.

12. $x^2 + ax + 2a + 2x$ **13.** $mn - 3n + 2m - 6$ 12. _____

                                                                                  13. _____

14. $z^3 + 4z^2 + 3z + 12$ **15.** $x^3 - x^2 - 5x + 5$ 14. _____

                                                                                  15. _____

Sullivan/Struve/Mazzarella, *Elementary Algebra*, 2e

Five-Minute Warm-Up 6.2
Factoring Trinomials of the Form $x^2 + bx + c$

1. List all possible combinations of factors for the given product.

   (a) -12 **(b)** 36 1a. _____

                                                                                1b. _____

*In Problems 2 and 3, find **(a)** the sum of the numbers and **(b)** the product of the numbers.*

   2. -3 and -12 **3.** 8 and -2 2a. _____

                                                                                  2b. _____

                                                                                 3a. _____

                                                                                 3b. _____

4. Find two integers with the following properties.

   (a) sum of -5 and product of -36 4a. _____

   (b) sum of -5 and product of 6 4b. _____

   (c) sum of -4 and product of -5 4c. _____

   (d) sum of 9 and product of 18 4d. _____

5. Determine the coefficients of $-x^2 + 7x - 2$.   5. _____

6. Find the product: $(x-6)(x-4)$ 6 _____

Guided Practice 6.2
Factoring Trinomials of the Form $x^2 + bx + c$

Objective 1: Factor Trinomials of the Form $x^2 + bx + c$

Step 1: Find the pair of integers whose product is c and whose sum is b. That is, determine m and n such that $mn = c$ and $m + n = b$.

Step 2: Write $x^2 + bx + c = (x + m)(x + n)$.

Step 3: Check your work by multiplying out the factored form.

1. Factor: $y^2 + 11y + 18$ *(See textbook Example 1)*

Step 1: We are looking for factors of $c = 18$ whose sum is $b = 11$. We begin by listing all factors of 18 and computing the sum of these factors.

(a)

Factors whose product is 18	1, 18	2, 9	3, 6	−1, −18	−2, -9	−3, −6
Sum of factors						

Which two factors sum to 11 and multiply to 18?

(b) _____

Step 2: We write the trinomial in the form $(y + m)(y + n)$.

(c) $y^2 + 11y + 18 =$ _____

Step 3: Check We multiply to verify our solution.

We leave it to you to verify the factorization.

2. Factor: $z^2 - 15z + 36$ *(See textbook Example 2)*

(a) Make a list of factors.

Factors whose product is 36	1, 36	2, 18	3, 12	4, 9	−1, −36	−2, −18	−3, −12	−4, −9
Sum of factors								

(b) Use the two factors whose sum is −15 to write the factorization of $z^2 - 15z + 36$.

2b. _____

3. Factor: $x^2 + x - 56$ *(See textbook Example 3)*

(a) Make a list of factors.

Factors whose product is −56							
Sum of factors							

(b) Use the two factors whose sum is 1 to write the factorization of $x^2 + x - 56$.

3b. _____

4. Factor: $n^2 - 2n - 24$ *(See textbook Example 4)*

(a) Make a list of factors.

Factors whose product is −24							
Sum of factors							

(b) Use the two factors whose sum is −2 to write the factorization of $n^2 - 2n - 24$.

4b. _____

5. Show that $x^2 + 3x - 6$ is prime. *(See textbook Example 5)*

Objective 2: Factor Out the GCF, Then Factor $x^2 + bx + c$

6. Factor: $3p^3 - 9p^2 - 54p$ *(See textbook Example 8)*

6. _____

Do the Math Exercises 6.2
Factoring Trinomials of the Form $x^2 + bx + c$

In Problems 1–5, factor each trinomial completely. If the trinomial cannot be factored, say it is prime.

1. $n^2 + 12n + 20$ **2.** $y^2 - 8y - 9$ 1. _____

       2. _____

3. $y^2 + 6y - 40$ **4.** $t^2 + 2t - 38$ 3. _____

       4. _____

5. $x^2 - 14xy + 24y^2$ 5. _____

In Problems 6–9, factor each trinomial completely by factoring out the GCF first and then factoring the resulting trinomial.

6. $4p^4 - 4p^3 - 8p^2$ **7.** $30x - 2x^2 - 100$ 6. _____

       7. _____

8. $-3x^2 - 18x - 15$

9. $-75x + x^3 + 10x^2$

8. _____

9. _____

In Problems 10–15, factor each polynomial completely. If the polynomial cannot be factored, say that it is prime.

10. $-2z^3 - 2z^2 + 24z$

11. $-r^2 - 12r - 36$

10. _____

11. _____

12. $x^2 - x + 6$

13. $-16x + x^3 - 6x^2$

12. _____

13. _____

14. $x^2 + 7xy + 12y^2$

15. $25 + 10x + x^2$

14. _____

15. _____

Five-Minute Warm-Up 6.3
Factoring Trinomials of the Form $ax^2 + bx + c$, $a \neq 1$

1. Determine the coefficients of $4n^2 + n - 1$?   1. _____

2. Write as the product of prime numbers.
   (a) 18   (b) 150   2a. _____

                                                                     2b. _____

3. Find the product: $(2x + 1)(3x - 2)$ 3. _____

4. Factor by grouping: $4x - 4y + ax - ay$ 4. _____

5. Factor out the GCF: $12r^3s^2 + 3rs - 6rs^4$ 5. _____

6. Factor out the greatest common binomial factor: $7z(2z + 1) - 4(2z + 1)$ 6. _____

Guided Practice 6.3
Factoring Trinomials of the Form $ax^2 + bx + c$, $a \neq 1$

Objective 1: Factor $ax^2 + bx + c$, $a \neq 1$ Using Trial and Error

1. Factor using trial and error: $3x^2 + 7x + 2$ *(See textbook Example 1)*

Step 1: List the possibilities for the first terms of each binomial whose product is ax^2.

(a) We list the possible ways of representing the first term, $3x^2$. Since 3 is a prime number, we have only one possiblity:

$$(\underline{\quad} + ?)(\underline{\quad} + ?)$$

Step 2: List the possibilities for the last terms of each binomial whose product is c.

(b) The last term, 2, is also prime. List the choices of pairs of factors that multiply to 2.

_____   _____

(c) However, did you notice that the coefficient of x, 7, is positive? To produce a positive sum, $7x$, we must have two positive factors. What is the only option for factors of 2?

_____

Step 3: Write out all the combinations of factors found in Steps 1 and 2. Multiply the binomials until a product is found that equals the trinomial.

(d) The order of the terms will change the product when the leading coefficient is not one. Once testing the possibilities produces the required product, write the factorization:

$$3x^2 + 7x + 2 = \underline{\hspace{4cm}}$$

You may continue to practice the trial and error method following textbook Examples 2–4. Do the Quick Check exercises after each textbook example and see if you can find the factorization easily. If it becomes too frustrating, move on to the second method for factoring trinomials whose leading coefficient is not one.

2. Factor using trial and error: $18a^2 - 35ab + 12b^2$ *(See textbook Example 5)*   2. _____

3. Factor: $-6x^2 + 17x + 14$ *(See textbook Example 6)*   3. _____

Objective 2: Factor $ax^2 + bx + c$, $a \neq 1$ Using Grouping

4. Factor by grouping: $3x^2 - 13x + 12$ *(See textbook Example 7)*

Step 1: Find the value of *ac*.

(a) Identify the coefficients: $a = \underline{\hspace{1cm}}$, $c = \underline{\hspace{1cm}}$

(b) The value of $a \bullet c = \underline{\hspace{1.5cm}}$

Step 2: Find the pair of integers, *m* and *n*, whose product is *ac* and whose sum is *b*.

(c)

Factors whose product is 36	−1, −36	−2, −18	−3, −12	−4, −9	−6, −6
Sum of factors					

(d) Which two factors multiply to 36 and sum to -13? $\underline{\hspace{3cm}}$

Step 3: Write
$$ax^2 + bx + c = ax^2 + mx + nx + c$$

(e) $3x^2 - 13x + 12 = \underline{\hspace{5cm}}$

Step 4: Factor by expression in Step 3 by grouping.

(f) $\underline{\hspace{7cm}}$

Step 5: Check

Multiply to verify that the factorization is correct.

5. Factor by grouping: $13x^2 + 9x - 4$ *(See textbook Example 8)*   5. $\underline{\hspace{2cm}}$

Name:
Instructor:

Date:
Section:

Do the Math Exercises 6.3
Factoring Trinomials of the Form $ax^2 + bx + c$, $a \neq 1$

In Problems 1–6, factor each polynomial completely using the trial and error method. Hint: None of the polynomials is prime.

1. $3x^2 + 16x - 12$

2. $5x^2 + 16x + 3$

1. _____

2. _____

3. $11p^2 - 46p + 8$

4. $3x^2 + 7xy + 2y^2$

3. _____

4. _____

5. $6x^2 - 14xy - 12y^2$

6. $-6x^2 - 3x + 45$

5. _____

6. _____

In Problems 7–12, factor each polynomial completely using the grouping method. Hint: None of the polynomials is prime.

7. $7n^2 - 27n - 4$

8. $25t^2 + 5t - 2$

7. _____

8. _____

9. $20t^2 + 21t + 4$

10. $12p^2 - 23p + 5$

9. _____

10. _____

11. $18x^2 + 6xy - 4y^2$

12. $-10y^2 + 47y + 15$

11. _____

12. _____

In Problems 13–15, factor completely. If a polynomial cannot be factored, say it is prime.

13. $18x^2 + 88x - 10$

14. $9x^3y + 6x^2y + 3xy$

13. _____

14. _____

15. $8x^2 + 14x - 7$

15. _____

Five-Minute Warm-Up 6.4
Factoring Special Products

1. Evaluate: $(-10)^2$ 1. _____

2. Evaluate: $(-3)^3$ 2. _____

3. Find the product: $(4x^2)^3$ 3. _____

4. Find the product: $(2p + 5)^2$ 4. _____

5. Find the product: $(4a + 9b)(4a - 9b)$ 5. _____

6. List the perfect squares that are less than 100. That is, find $1^2, 2^2, 3^2$ etc.

_____

7. List the perfect cubes that are less than 100. That is, find $1^3, 2^3, 3^3$ etc.

_____

Guided Practice 6.4
Factoring Special Products

Objective 1: Factor Perfect Square Trinomials

1. Find the product:

(a) $(A + B)^2 =$ _____

(b) $(A - B)^2 =$ _____

2. Factor completely: $z^2 - 8z + 16$. *(See textbook Example 1)*

Step 1: Determine whether the first term and the third term are perfect squares.

(a) What is the first term? _____ To be a perfect square, the first term must be of the form $(A)^2$. In this case case, $A =$ _____.

(b) What is the third term? _____ To be a perfect square, the third term must be of the form $(B)^2$. In this case case, $B =$ _____.

Step 2: Determine whether the middle term is 2 times or –2 times the product of the expression being squared in the first and last term.

(c) What is the middle term? _____ Is this equal to the product of 2 or -2 times A times B? _____

Step 3: Use $A^2 - 2AB + B^2 = (A - B)^2$

(d) $z^2 - 8z + 16 =$ _____

3. Factor each perfect square trinomial. *(See textbook Examples 2 and 4)*

(a) $p^2 + 22p + 121$

(b) $12m^2n - 36mn^2 + 27n^3$

3a. _____

3b. _____

Objective 2: Factor the Difference of Two Squares

4. Find the product: $(A - B)(A + B) =$ _____

5. Factor each difference of two squares completely. *(See textbook Examples 5 – 7)*

(a) $x^2 - 49$

(b) $25m^6 - 36n^4$

5a. _____

5b. _____

(c) $a^8 - 256$

(d) $12x^2y - 75y^3$

5c. _____

5d. _____

Objective 3: Factor the Sum or Difference of Two Cubes

6. Find the product:

(a) $(A + B)(A^2 - AB + B^2) =$ _____

(b) $(A - B)(A^2 + AB + B^2) =$ _____

(c) $(A + B)^3 =$ _____

7. Factor the sum or difference of two cubes. *(See textbook Examples 8 and 9)*

(a) $p^3 - 64$ **(b)** $8x^6 + 27y^3$ 7a. _____

7b. _____

8. Factor completely. *(See textbook Example 10)*

(a) $24x^4 + 375x$ **(b)** $250x^4y - 16xy^4$ 8a. _____

8b. _____

Do the Math Exercises 6.4
Factoring Special Products

In Problems 1–4, factor each perfect square trinomial completely.

1. $m^2 + 12m + 36$

2. $9a^2 - 12a + 4$

1. _____

2. _____

3. $16y^2 - 72y + 81$

4. $4a^2 + 20ab + 25b^2$

3. _____

4. _____

In Problems 5–6, factor each difference of two squares completely.

5. $36m^2 - 25n^2$

6. $a^4 - 16$

5. _____

6. _____

In Problems 7–10, factor each sum or difference of two cubes completely.

7. $64r^3 - 125s^3$

8. $m^9 - 27n^6$

7. _____

8. _____

9. $125y^3 + 27z^6$

10. $40x^3 + 135y^6$

9. _____

10. _____

In Problems 11–15, factor completely. If the polynomial is prime, state so.

11. $12n^3 - 36n^2 + 27n$

12. $32a^3 + 4b^6$

11. _____

12. _____

13. $x^4 - 225x^2$

14. $12m^2 - 14m + 21$

13. _____

14. _____

15. $3x^2 + 18x + 27$

15. _____

Sullivan/Struve/Mazzarella, *Elementary Algebra*, 2e

Five-Minute Warm-Up 6.5
Summary of Factoring Techniques

In Problems 1 – 2, find each product.

1. $(4x + 3y)(5x - 2y)$

1. _____

2. $(a + 4b)^2$

2. _____

In Problems 3 – 5, factor completely.

3. $-27a^3 + 9a^2 - 18a$

3. _____

4. $6m(2n - 1) - 9(2n - 1)$

4. _____

5. $4p^2 - 8pq + 3p - 6q$

5. _____

Guided Practice 6.5
Summary of Factoring Techniques

Objective 1: Factor Polynomials Completely

1. Review the **Steps for Factoring** listed at the beginning of textbook Section 6.5. Step 1 should always be

_____, if possible.

2. Factor: $4x^2 + 16x - 84$ *(See textbook Example 1)*

Step 1: Factor out the greatest common factor (GCF), if any exists.	Determine the GCF:	**(a)** GCF = _____
	Factor out the GCF:	**(b)** $4x^2 + 16x - 84 =$
		_____
Step 2: Identify the number of terms in the polynomial in parentheses.	How many terms are in the polynomial in parentheses?	**(c)** _____
Step 3: We concentrate on the trinomial in parentheses. It is not a perfect square trinomial. The trinomial has a leading coefficient of 1 so we try $(x + m)(x + n)$, where $mn = c$ and $m + n = b$.	Factor the trinomial:	**(d)** _____
Step 4: Check		We leave it to you to multiply the binomials and then distribute to verify that the factorization is correct.

3. Factor: $64a^2 - 81b^2$ *(See textbook Example 2)*

Step 1: Factor out the greatest common factor (GCF), if any exists.	There is no GCF.	
Step 2: Identify the number of terms in the polynomial.	How many terms are in the polynomial?	**(a)** _____
Step 3: Because the first term $64a^2 = (8a)^2$, and the second term, $81b^2 = (9b)^2$, are both perfect squares, we have the difference of two squares.	Factor the binomial:	**(b)** _____
Step 4: Check		We leave it to you to verify that the factorization is correct.

4. Factor: $8a^2 + 24ab + 18b^2$ *(See textbook Example 3)*

Step 1: Factor out the greatest common factor (GCF), if any exists.

What is the GCF of 8, 24, and 18?

(a) _____

Factor out the GCF:

(b) _____

Step 2: Identify the number of terms in the polynomial in parentheses.

How many terms are in the polynomial?

(c) _____

Step 3: We concentrate on the polynomial in parentheses. Is it a perfect square trinomial?

(d) What is the first term? _____ To be a perfect square, the first term must be of the form $(A)^2$. In this case case, $A =$ _____.

(e) What is the third term? _____ To be a perfect square, the thrid term must be of the form $(B)^2$. In this case case, $B =$ ____.

(f) What is the middle term? _____ Is this equal to the product of 2 or -2 times A times B? _____

Use $A^2 - 2AB + B^2 = (A - B)^2$

Write the factorization:

(g) _____

Step 4: Check

We leave it to you to verify that the factorization is correct.

5. Factor completely: $27p^6 + 1$ *(See textbook Example 4)*

5. _____

6. Factor completely: $2n^3 - 10n^2 - 6n + 30$ *(See textbook Example 6)*

6. _____

7. Factor completely: $-3xy^2 - 12xy - 9x$ *(See textbook Example 7)*

7. _____

Do the Math Exercises 6.5
Summary of Factoring Techniques

In Problems 1–15, factor completely. If a polynomial cannot be factored, say it is prime.

1. $x^2 - 256$

2. $1 - y^9$

1. _____

2. _____

3. $x^2 + 5x - 6$

4. $2x^3 - x^2 - 18x + 9$

3. _____

4. _____

5. $100x^2 - 25y^2$

6. $6s^2t^2 + st - 1$

5. _____

6. _____

7. $8x^6 + 125y^3$

8. $48m - 3m^9$

7. _____

8. _____

9. $4t^4 + 16t^2$

10. $8n^3 - 18n$

9. _____

10. _____

11. $n^2 + 2n + 8$

12. $24p^3q + 81q^4$

11. _____

12. _____

13. $x^2(x+y) - 16(x+y)$

14. $-9x - 3x^3 - 12x^2$

13. _____

14. _____

15. $6a^3 - 5a^2b + ab^2$

15. _____

Sullivan/Struve/Mazzarella, *Elementary Algebra,* 2e

Five-Minute Warm-Up 6.6
Solving Polynomial Equations by Factoring

1. Solve: $2x - 1 = 0$ 1. _____

2. Solve: $-3(x - 2) + 15 = 0$ 2. _____

3. Evaluate $3n^2 + 4n - 5$ for
  (a) $n = 2$ **(b)** $n = -3$ 3a. _____

                    3b. _____

In Problems 4 – 7, factor completely.

4. $x^2 + 2x - 63$ 4. _____

5. $4p^2 + 4p - 3$ 5. _____

6. $x^2 - 81$ 6. _____

7. $n^2 - 16n + 64$ 7. _____

Guided Practice 6.6
Solving Polynomial Equations by Factoring

Objective 1: Solve Quadratic Equations Using the Zero-Product Property

1. State the **Zero-Product Property**. _____

_____

_____

2. Solve $(x + 1)(2x - 3) = 0$ *(See textbook Example 1)*   2. _____

3. A second-degree equation is also called a _____

4. What does it mean to write a quadratic equation in *standard form*? _____

_____

5. Solve: $x^2 + 5x - 24 = 0$ *(See textbook Example 2)*

Step 1: Is the quadratic equation in standard form? Yes, it is written in the form $ax^2 + bx + c = 0$.	$x^2 + 5x - 24 = 0$
Step 2: Factor the expression on the left side of the equation.	**(a)** _____
Step 3: Set each factor to 0. **(b)** _____	**(c)** _____
Step 4: Solve each first-degree equation. **(d)** _____	**(e)** _____
Step 5: Check Substitute your values into the original equation.	We leave it to you to verify the solutions.
Write the solution set:	**(f)** _____

6. Solve: $3t^2 + 11t = 4$ *(See textbook Example 3)*

   (a) Write the quadratic equation in standard form: _____

   (b) Factor the polynomial on the left side of the equation: _____

   (c) Solve and state the solution set. _____

7. Solve: $(x-2)(x-3) = 56$ *(See textbook Example 5)*

    (a) To solve this equation, the first step is _____

    (b) Next, write the quadratic equation in _____

    (c) Factor the polynomial on the left side of the equation: _____

    (d) Solve and state the solution set. _____

8. Solve: $4x^2 + 36 = 24x$ *(See textbook Examples 6 and 7)*

    (a) Write the quadratic equation in standard form: _____

    (b) Factor the polynomial on the left side of the equation: _____

    (c) Solve and state the solution set. _____

Objective 2: Solve Polynomial Equations of Degree Three or Higher Using the Zero-Product Property

9. Solve: $p^3 + 2p^2 - 9p - 18 = 0$ *(See textbook Example 9)*

Step 1: Write the equation in standard form.	The polynomial is already in standard form.	$p^3 + 2p^2 - 9p - 18 = 0$	
Step 2: Factor the expression on the left side of the equation. Begin with the GCF, if any.	Does the equation contain a GCF?	**(a)** _____	
	How many terms are in the polynomial?	**(b)** _____	
	What technique should you use?	**(c)** _____	
	Factor the polynomial:	**(d)** _____	
Step 3: Set each factor to 0.	**(e)** _____	**(f)** _____	**(g)** _____
Step 4: Solve each first-degree equation.	**(h)** _____	**(i)** _____	**(j)** _____
Step 5: Check Substitute your values into the original equation.		We leave it to you to verify the solutions.	
	Write the solution set:	**(k)** _____	

Sullivan/Struve/Mazzarella, *Elementary Algebra*, 2e

Do the Math Exercises 6.6
Solving Polynomial Equations by Factoring

In Problems 1–2, solve each equation using the Zero-Product Property.

1. $3x(x + 9) = 0$

2. $(4z - 3)(z + 4) = 0$

1. _____

2. _____

In Problems 3–4, identify each equation as a linear equation or a quadratic equation.

3. $2x + 1 - (x + 7) = 3x + 1$

4. $(x + 2)(x - 2) = 14$

3. _____

4. _____

In Problems 5–12, solve each quadratic equation by factoring.

5. $p^2 - 5p - 24 = 0$

6. $14x - 49x^2 = 0$

5. _____

6. _____

7. $3x^2 + x - 14 = 0$

8. $k^2 + 12k + 36 = 0$

7. _____

8. _____

9. $a^2 - 6a = 16$

10. $m^2 - 30 = 7m$

9. _____

10. _____

11. $4p - 3 = -4p^2$

12. $(x + 5)(x - 3) = 9$

11. _____

12. _____

In Problems 13–14, solve each polynomial equation by factoring.

13. $3x^3 + x^2 - 14x = 0$

14. $m^3 + 2m^2 - 9m - 18 = 0$

13. _____

14. _____

15. Tossing a Ball A ball is thrown vertically upward from the ground with an initial velocity of 64 feet per second. Solve the equation $-16t^2 + 64t = 48$ to find the time t (in seconds) at which the ball is 48 feet from the ground.

15. _____

Five-Minute Warm-Up 6.7
Modeling and Solving Problems with Quadratic Equations

1. Solve: $x(x+4) = 45$

    1. _____

2. Solve: $p^2 + 12p + 36 = 0$

    2. _____

3. Solve: $6x^2 + 17x + 5 = 0$

    3. _____

4. The lengths of the two legs of a right triangle are 9 and 12 inches. Find the length of the hypotenuse.

    4. _____

5. Find the product: $(x-5)^2$

    5. _____

6. Factor out the GCF: $-18h^3 + 27h^2 - 9h$

    6. _____

Guided Practice 6.7
Modeling and Solving Problems with Quadratic Equations

Objective 1: Model and Solve Problems Involving Quadratic Equations

1. A ball is thrown off a cliff from a height of 64 feet above sea level. The height h of the ball above the water (in feet) at any time t (in seconds) can be modeled by the equation

$$h = -16t^2 + 48t + 64$$

(See textbook Example 1)

    (a) When will the height of the ball be 64 feet above sea level?   1a. _____

    (b) When will the ball strike the water?   1b. _____

2. The length and width of two sides of a rectangle are consecutive odd integers. The area of the rectangle is 255 square centimeters. Find the dimensions of the rectangle. *(See textbook Example 2)*

    Step 1: Identify This is a geometry problem involving the area of a rectangle. It also involves consecutive odd integers.

    (a) If n represents one of the odd integers, express the next consecutive odd integer: 2a. _____

    (b) In general, what formula do we use to calculate the area of a rectangle?   2b. _____

    (c) Step 2: Name Use the variables from (a) to identify an expression that   2c.
    represents the length and width of the rectangle.   width = _____

            length = _____

    (d) Step 3: Translate Write an equation that will model the area of this rectangle.   2d. _____

    (e) Step 4: Solve the equation from step 3.   2e. _____

    Step 5: Check Is your answer reasonable? Does it meet the necessary conditions?

    (f) Step 6: Answer the question. _____

Guided Practice 6.7

Objective 2: Model and Solve Problems Using the Pythagorean Theorem

3. In your own words, state the Pythagorean Theorem.

4. If x and y are the lengths of the legs and z is the length of the hypotenuse, write an equation which uses these variables to state the Pythagorean Theorem.

4. _____

5. Use the Pythagorean Theorem to find the missing length in the right triangle. *(See textbook Example 4)*

   (a) one leg = 18 cm; second leg = 24 cm, find the hypotenuse.

   (b) one leg is 24 ft; hypotenuse is 25 ft, find the second leg.

5a. _____

5b. _____

6. In a right triangle, the hypotenuse measures $(2x - 3)$ meters. If one leg measures x meters and the other leg measures $(x + 3)$ meters,

   (a) solve for x.

6a. _____

   (b) What are the lengths of the sides of the right triangle?

6b. _____

Do the Math Exercises 6.7
Modeling and Solving Problems with Quadratic Equations

In Problems 1–2, use the given area to find the missing sides of the rectangle.

1. $A = 250$

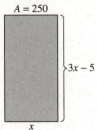

$3x - 5$

x

2. $A = 364$

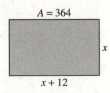

x

$x + 12$

1. _____

2. _____

In Problems 3–4, use the given area to find the height and base of the triangle.

3. $A = 42$

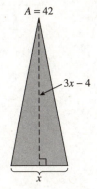

$3x - 4$

x

4. $A = 84$

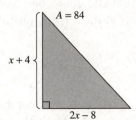

$x + 4$

$2x - 8$

3. _____

4. _____

In Problems 5–6, use the given area to find the dimensions of the quadrilateral.

5. $A = 77$

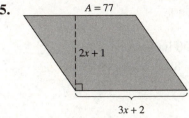

$2x + 1$

$3x + 2$

6. $2x + 1$

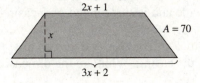

x

$A = 70$

$3x + 2$

5. _____

6. _____

In Problems 7–8, use the Pythagorean Theorem to find the lengths of the sides of the triangle.

7.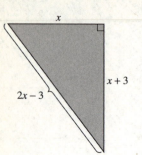

8.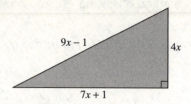

7. _____

8. _____

9. **Projectile Motion** If $h = -16t^2 + 96t$ represents the height of a rocket, in feet, t seconds after it was fired, when will the rocket be 144 feet high?

9. _____

10. **Rectangular Room** The width of a rectangular room is 4 feet less than the length. If the area of the room is 21 square feet, find the dimensions of the room.

10. _____

11. **Triangle** The base of a triangle is 2 meters more than the height. If the area of the triangle is 24 square meters, find the base and height of the triangle.

11. _____

12. **Playing Field Dimensions** The length of a rectangular playing field is 3 yards less than twice the width. If the area of the field is 104 square yards, what are its dimensions?

12. _____

13. **Jasper's Big-Screen TV** Jasper purchased a new big-screen TV. The TV screen is 10 inches taller than it is wide and is surrounded by a casing that is 2 inches wide.

(a) Jasper lost his tape measure but sees on the box that the TV measures 50 inches on the diagonal, including the casing. What are the dimensions of the TV screen?

13a. _____

(b) Jasper knows that the size of the opening where he wants the TV installed is 29 by 42 inches. Will this TV fit into his space?

13b. _____

Sullivan/Struve/Mazzarella, *Elementary Algebra*, 2e

Name: \
Instructor:

Date: \
Section:

Five-Minute Warm-Up 7.1
Simplifying Rational Expressions

1. Evaluate $\dfrac{-4a + 2b}{3}$ for $a = -2$ and $b = -7$.

1. _____

2. Factor: $4x^2 + x - 3$

2. _____

3. Solve: $2x^2 - x - 15 = 0$

3. _____

4. Write $\dfrac{30}{63}$ in lowest terms.

4. _____

5. Divide: $\dfrac{24x^2 y}{36xy^3}$

5. _____

Guided Practice 7.1
Simplifying Rational Expressions

Objective 1: Evaluate a Rational Expression

1. Evaluate the following rational expressions. *(See textbook Example 1)*

(a) $\dfrac{3}{2x+4}$ for $x=3$ (b) $\dfrac{5x-3}{x+1}$ for $x=7$ 1a. _____

1b. _____

2. Evaluate the following rational expressions. *(See textbook Examples 2 and 3)*

(a) $\dfrac{x+9}{x^2+8x-12}$ for $x=-3$ (b) $\dfrac{-3a+2b}{-a-b}$ for $a=6$ and $b=2$ 2a. _____

2b. _____

Objective 2: Determine Values for which a Rational Expression is Undefined

3. To determine the undefined values of a rational expression, find the values where _____.

4. Find the value(s) of x for which the expression is undefined. *(See textbook Examples 4 and 5)*

(a) $\dfrac{x-3}{x+8}$ (b) $\dfrac{x^2-2x-3}{x^2-9x+18}$ 4a. _____

4b. _____

Objective 3: Simplify Rational Expressions

5. To simplify a rational expression always use the following steps:

(a) Completely _____ the numerator and denominator of the rational expression.

(b) _____ _____ the common factors.

6. Simplify: $\dfrac{6x - 18}{x^2 - 5x + 6}$, $x \neq 2$, $x \neq 3$ *(See textbook Example 6)*

Step 1: Completely factor the numerator and the denominator.	This problem has included the restrictions on the variable. These excluded values cause the denominator to equal zero and will not affect our work.
	Factor each expression: **(a)** $\dfrac{6x - 18}{x^2 - 5x + 6} = $ _____
Step 2: Divide out common factors. The rational expression is completely simplified if the numerator and the denominator share no common factor other than 1.	Divide out common factors: **(b)** _____
	Write the simplified expression: **(c)** _____

7. Simplify: $\dfrac{2x^2 + 12x + 10}{4x^2 + 28x + 40}$ *(See textbook Example 8)*

7. _____

8. Simplify: $\dfrac{50 - 2x^2}{4x^2 + 28x + 40}$ *(See textbook Example 9)*

8. _____

Do the Math Exercises 7.1
Simplifying Rational Expressions

In Problems 1–4, evaluate each expression for the given values.

1. $\dfrac{x}{x+4}$
   (a) $x = 8$
   (b) $x = -4$
   (c) $x = 0$

2. $\dfrac{2m-1}{m}$
   (a) $m = 4$
   (b) $m = 1$
   (c) $m = -1$

1a. _____
1b. _____
1c. _____

2a. _____
2b. _____
2c. _____

3. $\dfrac{a^2 - 2a}{a-4}$
   (a) $a = 5$
   (b) $a = -1$
   (c) $a = -4$

4. $\dfrac{b^2 - a^2}{(a-b)^2}$
   (a) $a = 3, b = 2$
   (b) $a = 5, b = 4$
   (c) $a = -2, b = 2$

3a. _____
3b. _____
3c. _____

4a. _____
4b. _____
4c. _____

In Problems 5–8, find the value(s) of the variable for which the rational expression is undefined.

5. $\dfrac{5m^3}{m+8}$

6. $\dfrac{12}{4a-3}$

5. _____
6. _____

7. $\dfrac{2x^2}{x^2 + x - 2}$

8. $\dfrac{3h+2}{h^3 + 5h^2 + 4h}$

7. _____
8. _____

In Problems 9–16, simplify each rational expression. Assume that no variable has a value which results in a denominator with a value of zero.

9. $\dfrac{6p^2 + 3p}{3p}$

10. $\dfrac{4 - z}{z - 4}$

9. _____

10. _____

11. $\dfrac{x^2 - 9}{x^2 + 5x + 6}$

12. $\dfrac{x^3 + 3x^2 + 2x}{x^3 - 2x^2 - 3x}$

11. _____

12. _____

13. $\dfrac{a - b}{b^2 - a^2}$

14. $\dfrac{z^3 + 3z^2 + 2z + 6}{z^2 + 5z + 6}$

13. _____

14. _____

15. $\dfrac{2x^2 + 5x - 3}{4x^2 - 8x + 3}$

16. $\dfrac{5 + 4x - x^2}{x^2 - 25}$

15. _____

16. _____

Sullivan/Struve/Mazzarella, *Elementary Algebra*, 2e

Five-Minute Warm-Up 7.2
Multiplying and Dividing Rational Expressions

1. Find the product: $\dfrac{21}{4} \bullet \dfrac{18}{49}$ 1. _____

2. Find the reciprocal of $-\dfrac{5}{2}$.   2. _____

3. Find the quotient: $\dfrac{15}{36} \div \dfrac{70}{24}$ 3. _____

4. Factor: $-5x^2 + 25x$ 4. _____

5. Simplify the rational expression: $\dfrac{x^2 - 9}{x^2 - 3x - 18}$ 5. _____

Guided Practice 7.2
Multiplying and Dividing Rational Expressions

Objective 1: Multiply Rational Expressions

1. Multiply $\dfrac{x+3}{8} \cdot \dfrac{4x-12}{x^2-9}$. Simplify the result, if possible. *(See textbook Example 1)*

Step 1: Completely factor the polynomials in each numerator and denominator.

$$\dfrac{x+3}{8} \cdot \dfrac{4x-12}{x^2-9} =$$

Factor the numerator and denominator:

(a) _____

Step 2: Multiply using $\dfrac{a}{b} \cdot \dfrac{c}{d} = \dfrac{ac}{bd}$.

Write the product:

(b) _____

Step 3: Divide out common factors in the numerator and denominator.

Divide out common factors and express the answer as a simplified rational expression in factored form. **(c)** _____

Write the simplified result:

(d) _____

2. Multiply $\dfrac{x^2-4x}{x^2-4} \cdot \dfrac{x^2-x-6}{x-4}$. Simplify the result, is possible. *(See textbook Example 2)*

(a) Completely factor the polynomials in each numerator and denominator.

(a) _____

(b) Multiply.

(b) _____

(c) Divide out common factors in the numerator and denominator.

(c) _____

(d) Express the answer as a simplified rational expression in factored form.

(d) _____

3. Find the product and simplify: $\dfrac{x^2-y^2}{13x^2-13xy} \cdot \dfrac{26y-26x}{x^2+2xy+y^2}$ *(See textbook Example 4)* 3. _____

Objective 2: Divide Rational Expressions

4. Divide $\dfrac{3x+15}{36} \div \dfrac{5x+25}{24}$. Simplify the result, is possible. *(See textbook Example 5)*

Step 1: Multiply the dividend by the reciprocal of the divisor.

$$\dfrac{3x+15}{36} \div \dfrac{5x+25}{24} =$$

Rewrite the quotient as a product: **(a)** _____

Step 2: Completely factor the polynomials in each numerator and denominator.

Factor each expression: **(b)** _____

Step 3: Multiply.

Write the product: **(c)** _____

Step 4: Divide out common factors in the numerator and denominator.

Divide out the common factors: **(d)** _____

Express the answer as a simplified rational expression in factored form: **(e)** _____

5. Find the quotient and simplify: $\dfrac{x^2-4x+3}{2x^2-5x+3} \div (x-3)$ *(See textbook Example 7)*   5. _____

6. Find the quotient and simplify: $\dfrac{\dfrac{3y^2-3x^2}{x^2+2xy+y^2}}{\dfrac{6x^2-6xy}{3y+3x}}$ *(See textbook Example 8)*   6. _____

Do the Math Exercises 7.2
Multiplying and Dividing Rational Expressions

In Problems 1–4, multiply.

1. $\dfrac{x-4}{4} \cdot \dfrac{12x}{x^2-16}$

2. $\dfrac{8n-8}{n^2-3n+2} \cdot \dfrac{n+2}{12}$

1. _____

2. _____

3. $\dfrac{z^2-4}{3z-2} \cdot \dfrac{3z^2+7z-6}{z^2+z-6}$

4. $(x-3) \cdot \dfrac{x+2}{x^2-5x+6}$

3. _____

4. _____

In Problems 5–8, divide.

5. $\dfrac{p+7}{3p+21} \div \dfrac{p}{6}$

6. $\dfrac{z^2-25}{10z} \div \dfrac{z^2-10z+25}{5z}$

5. _____

6. _____

7. $\dfrac{3y^2+6y}{8} \div \dfrac{y+2}{12y-12}$

8. $\dfrac{4z^2+12z+9}{8z+16} \div \dfrac{(2z+3)^2}{12z+24}$

7. _____

8. _____

In Problems 9–16, perform the indicated operation.

9. $\dfrac{x-4}{12x-18} \cdot \dfrac{6}{x^2-16}$

10. $\dfrac{x+5}{3} \div \dfrac{30x}{4x+20}$

9. _____

10. _____

11. $\dfrac{2r^2+rs-3s^2}{r^2-s^2} \cdot \dfrac{r^2-2rs-3s^2}{2r+3s}$

12. $\dfrac{2x^2-5x+3}{x^2-1} \cdot \dfrac{x^2+1}{2x^2-x-3}$

11. _____

12. _____

13. $\dfrac{(x-3)^2}{8xy^2} \div \dfrac{x^2-9}{(4x^2y)^2}$

14. $\dfrac{p^2-49}{p^2-5p-14} \cdot \dfrac{p-7}{14-5p-p^2}$

13. _____

14. _____

15. $\dfrac{1}{b^2+b-12} \div \dfrac{1}{b^2-5b-36}$

16. $\dfrac{\dfrac{t^2}{t^2-16}}{\dfrac{t^2-3t}{t^2-t-12}}$

15. _____

16. _____

Five-Minute Warm-Up 7.3
Adding and Subtracting Rational Expressions with a Common Denominator

1. Write $\dfrac{27}{45}$ in lowest terms.

1. _____

2. Find each sum and simplify:

   (a) $\dfrac{8}{3} + \dfrac{5}{3}$

   (b) $\dfrac{11}{8} + \dfrac{9}{8}$

2a. _____

2b. _____

3. Find each difference and simplify:

   (a) $\dfrac{9}{2} - \dfrac{7}{2}$

   (b) $\dfrac{13}{24} - \left(-\dfrac{2}{24}\right)$

3a. _____

3b. _____

4. Determine the additive inverse of $\dfrac{1}{2}$.

4. _____

5. Simplify: $-(2x - 3)$

5. _____

Guided Practice 7.3
Adding and Subtracting Rational Expressions with a Common Denominator

Objective 1: Add Rational Expressions with a Common Denominator

1. Find the sum and simplify, if possible: $\dfrac{2x}{x^2 - 3x} + \dfrac{5x}{x^2 - 3x}$ *(See textbook Example 1)*

Step 1: Add the numerators and write the result over the common denominator.	Use $\dfrac{a}{c} + \dfrac{b}{c} = \dfrac{a+b}{c}$:	**(a)** _____
Step 2: Simplify the rational expression.	Factor the numerator and denominator:	**(b)** _____
	Divide out the common factors:	**(c)** _____
	Express the answer in simplified form:	**(d)** _____

2. Find the sum and simplify, if possible: $\dfrac{3x - 10}{x^2 - 25} + \dfrac{2x - 15}{x^2 - 25}$ *(See textbook Example 3)*   2. _____

Objective 2: Subtract Rational Expressions with a Common Denominator

3. Find the difference and simplify, if possible: $\dfrac{x^2}{x^2 - 5x - 6} - \dfrac{1}{x^2 - 5x - 6}$ *(See textbook Example 4)*

Step 1: Subtract the numerators and write the result over the common denominator.	Use $\dfrac{a}{c} - \dfrac{b}{c} = \dfrac{a-b}{c}$:	**(a)** _____
Step 2: Simplify the rational expression.	Factor the numerator and denominator:	**(b)** _____
	Divide out the common factors:	**(c)** _____
	Express the answer in simplified form:	**(d)** _____

4. Find the difference and simplify if possible: $\dfrac{2x^2 + 3x}{x - 2} - \dfrac{x^2 + 3x + 4}{x - 2}$ *(See textbook*

4. _____

Example 5)

Objective 3: Add or Subtract Rational Expressions with Opposite Denominators

5. Find the sum and simplify, if possible: $\dfrac{6n - 5}{n - 5} + \dfrac{2n - n^2}{5 - n}$ *(See textbook Example 6)*

5. _____

6. Find the difference and simplify, if possible: $\dfrac{x^2 + 1}{x^2 - 9} - \dfrac{4x + 2}{9 - x^2}$ *(See textbook*

6. _____

Example 7)

Do the Math Exercises 7.3
Adding and Subtracting Rational Expressions with a Common Denominator

In Problems 1–4, add the rational expressions.

1. $\dfrac{4n+1}{8}+\dfrac{12n-1}{8}$

2. $\dfrac{4p-1}{p-1}+\dfrac{p+1}{p-1}$

1. _____

2. _____

3. $\dfrac{3x^2+4}{3x-6}+\dfrac{3x^2-28}{3x-6}$

4. $\dfrac{2x}{2x^2-x-15}+\dfrac{5}{2x^2-x-15}$

3. _____

4. _____

In Problems 5–8, subtract the rational expressions.

5. $\dfrac{4a^2}{3a-1}-\dfrac{a^2+a}{3a-1}$

6. $\dfrac{7x+13}{3x}-\dfrac{x+1}{3x}$

5. _____

6. _____

7. $\dfrac{2z^2+7z}{z^2-1}-\dfrac{z^2-6}{z^2-1}$

8. $\dfrac{2n^2+n}{n^2-n-2}-\dfrac{n^2+6}{n^2-n-2}$

7. _____

8. _____

In Problems 9–12, add or subtract the rational expressions.

9. $\dfrac{4}{x-1}+\dfrac{8x}{1-x}$

10. $\dfrac{x^2+3}{x-1}-\dfrac{x+3}{1-x}$

9. _____

10. _____

11. $\dfrac{x^2+6x}{x^2-1}+\dfrac{4x+3}{1-x^2}$

12. $\dfrac{m^2+2mn}{n^2-m^2}-\dfrac{n^2}{m^2-n^2}$

11. _____

12. _____

In Problems 13–16, perform the indicated operation.

13. $\dfrac{3n^2}{2n-1}-\dfrac{6n}{2n-1}$

14. $\dfrac{x^2}{x^2-9}+\dfrac{6x+9}{x^2-9}$

13. _____

14. _____

15. $\dfrac{2s-3t}{s-t}+\dfrac{6s-4t}{t-s}$

16. $\dfrac{n+3}{2n^2-n-3}-\dfrac{-n^2-3n}{2n^2-n-3}$

15. _____

16. _____

Five-Minute Warm-Up 7.4
Finding the Least Common Denominator and Forming Equivalent Rational Expressions

1. Write $\dfrac{7}{15}$ as a fraction with 45 as its denominator.   1.

2. Find the least common denominator (LCD) of $\dfrac{9}{20}$ and $\dfrac{2}{45}$.   2.

3. Write each of the rational expressions in Problem 2 as equivalent expressions using the   3. _____
LCD.

4. Identify the missing factor: $12x^2 y \bullet ? = 36x^2 y^3$ 4. _____

5. Factor: $x^2 - 4x + 4$ 5. _____

6. Factor: $-3x^2 - 9x$ 6. _____

Guided Practice 7.4
Finding the Least Common Denominator and Forming Equivalent Rational Expressions

Objective 1: Find the Least Common Denominator of Two or More Rational Expressions

1. Find the least common denominator (LCD) of $\dfrac{7}{24}$ and $\dfrac{3}{20}$. *(See textbook Example 1)*

Step 1: Write each denominator as the product of prime factors.

Write 24 as the product of prime factors: **(a)** _____

(b) _____

Write 20 as the product of prime factors:

Step 2: Write down the factor(s) that the denominators share, if any. Then write down the remaining factors the greatest number of times that the factor occurs in a denominator.

List the shared factors: **(c)** _____

List the remaining factors the greatest number of times that the factors occurs in a factorization: **(d)** _____

Step 3: Multiply the factors listed in Step 2. The product is the least common denominator (LCD).

Multiply the factors in (c) and (d): **(e)** LCD = _____

2. Find the least common denominator (LCD) of the rational expressions $\dfrac{7}{6x^2}$ and $\dfrac{5}{9x}$. *(See textbook Example 2)*

Step 1: Express each denominator as the product of factors. Write the factored form using exponents, if possible.

Write $6x^2$ as the product of prime factors: **(a)** _____

Write $9x$ as the product of prime factors: **(b)** _____

Step 2: List the common factors raised to the highest power. Then list the factors that are not common.

List the shared factors, using the highest power: **(c)** _____

List the remaining factors: **(d)** _____

Step 3: Multiply the factors listed in Step 2. The product is the least common denominator (LCD).

Multiply the factors in (c) and (d): **(e)** LCD = _____

3. Find the LCD of the rational expressions: $\dfrac{7c}{8a^3b}$ and $\dfrac{11d}{10a^2b^2}$ *(See textbook Example 3)*   3. _____

4. Find the LCD of the rational expressions: $\dfrac{8}{7a^2 - 14a}$ and $\dfrac{5}{2a^2}$ *(See textbook Example 4)* 4. _____

5. Find the LCD of the rational expressions: $\dfrac{4x}{x^2 - 9}$ and $\dfrac{x^2}{x^2 - 6x + 9}$ *(See textbook*

Example 5)   5. _____

Objective 2: *Write a Rational Expression That Is Equivalent to a Given Rational Expression*

6. Write the rational expression $\dfrac{x - 2}{3x^2 - x}$ as an equivalent rational expression with denominator $9x^3 - 3x^2$.

(See textbook Example 8)

Step 1: Write each denominator in factored form.	Write $3x^2 - x$ as the product of prime factors:	**(a)** _____
	Write $9x^3 - 3x^2$ as the product of prime factors:	**(b)** _____
Step 2: Determine the "missing factor(s)".	Compare (a) and (b). What is (are) the missing factor(s)?	**(c)** _____
Step 3: Multiply the original rational expression by 1.	Write 1 as $\dfrac{\text{missing factor}}{\text{missing factor}}$.	**(d)** _____
Step 4: Find the product. Leave the denominator in factored form.	Multiply the expression in (d) and the original rational expression. This product is the equivalent rational expression.	**(e)** _____

Objective 3: *Use the LCD to Write Equivalent Rational Expressions*

7. Find the LCD of the rational expressions $\dfrac{4}{9xy^2}$ and $\dfrac{5}{12x^2}$. Then rewrite each rational

expression with the LCD. *(See textbook Example 9)*

7. _____

_____

_____

Sullivan/Struve/Mazzarella, *Elementary Algebra*, 2e

Do the Math Exercises 7.4
Finding the Least Common Denominator and Forming Equivalent Rational Expressions

In Problems 1–6, find the LCD of the given rational expressions.

1. $\dfrac{3}{7y^2}; \dfrac{4}{49y}$

2. $\dfrac{2}{6x-18}; \dfrac{3}{8x-24}$

1. _____

2. _____

3. $\dfrac{5}{2x^2-12x+18}; \dfrac{4}{4x^2-36}$

4. $\dfrac{8}{x^2-1}; \dfrac{3}{x^2-2x+1}$

3. _____

4. _____

5. $\dfrac{-1}{c^2-49}; \dfrac{5}{21-3c}$

6. $\dfrac{3z}{(z+5)(z-4)}; \dfrac{-z}{(4-z)(5+z)}$

5. _____

6. _____

In Problems 7–10, write an equivalent rational expression with the given denominator.

7. $\dfrac{7a+1}{abc}$ with denominator a^2b^2c

8. $\dfrac{3}{x+2}$ with denominator x^2+5x+6

7. _____

8. _____

9. $\dfrac{7a}{3a^2+9a+6}$ with denominator $6a^2+18a+12$ **10.** 7 with denominator t^2+1

9. _____

10. _____

In Problems 11–18, find the LCD of the rational expressions. Then rewrite each as an equivalent rational expression with the LCD.

11. $\dfrac{4}{5n^2}; \dfrac{3}{7n}$

12. $\dfrac{2p^2-1}{6p}; \dfrac{3p^3+2}{8p^2}$

11. _____

12. _____

13. $\dfrac{x+2}{x}; \dfrac{x}{x+2}$

14. $\dfrac{3b}{7a}; \dfrac{2a}{7a+14b}$

13. _____

14. _____

15. $\dfrac{5}{m-6}; \dfrac{2m}{6-m}$

16. $\dfrac{4a}{a^2-1}; \dfrac{7}{a+1}$

15. _____

16. _____

17. $\dfrac{4n}{n^2-n-6}; \dfrac{2}{n^2+4n+4}$

18. $\dfrac{3}{2n^2-7n+3}; \dfrac{2n}{n-3}$

17. _____

18. _____

Sullivan/Struve/Mazzarella, *Elementary Algebra*, 2e

Five-Minute Warm-Up 7.5
Adding and Subtracting Rational Expressions with Unlike Denominators

1. Determine the least common denominator (LCD) and then write the rational expressions as equivalent expressions using the LCD.

   (a) $\dfrac{3}{2}$ and $\dfrac{5}{3}$

   (b) $\dfrac{7}{12}$ and $\dfrac{5}{28}$

   1a. _____

   1b. _____

2. Factor: $6z^2 - 7z + 2$

   2. _____

3. Simplify: $\dfrac{8x^2 - 4x}{4x}$

   3. _____

4. Simplify: $\dfrac{16 - x^2}{2x^2 - 8x}$

   4. _____

5. Find the sum: $\dfrac{3}{8x} + \dfrac{1}{8x}$

   5. _____

Guided Practice 7.5
Adding and Subtracting Rational Expressions with Unlike Denominators

Objective 1: Add and Subtract Rational Expressions with Unlike Denominators

1. Find the sum and simplify, if possible: $\dfrac{7}{15} + \dfrac{10}{25}$ *(See textbook Example 1)*

Step 1: Find the least common denominator of 15 and 25.

Write 15 as the product of prime factors: **(a)** _____

Write 25 as the product of prime factors: **(b)** _____

Determine the LCD: **(c)** _____

Step 2: Write each fraction as an equivalent fraction using the LCD.

Find the equivalent fraction: **(d)** $\dfrac{7}{15} \bullet \dfrac{\rule{1cm}{0.4pt}}{\rule{1cm}{0.4pt}} = \dfrac{\rule{1cm}{0.4pt}}{\rule{1cm}{0.4pt}}$

Find the equivalent fraction: **(e)** $\dfrac{10}{25} \bullet \dfrac{\rule{1cm}{0.4pt}}{\rule{1cm}{0.4pt}} = \dfrac{\rule{1cm}{0.4pt}}{\rule{1cm}{0.4pt}}$

Step 3: Find the sum of the numerators, written over the common denominator.

Use $\dfrac{a}{c} + \dfrac{b}{c} = \dfrac{a+b}{c}$:

(f) _____

Step 4: Simplify, if possible.

What is the sum, reduced to lowest terms?   **(g)** _____

2. Find the sum and simplify, if possible: $\dfrac{5}{12x} + \dfrac{7x}{18x^2}$ *(See textbook Example 2)*

Step 1: Find the least common denominator.

Write $12x$ as the product of prime factors: **(a)** _____

Write $18x^2$ as the product of prime factors: **(b)** _____

Determine the LCD: **(c)** _____

Step 2: Write each rational expression as an equivalent expression using the LCD.

Find the equivalent expression: **(d)** $\dfrac{5}{12x} \bullet \dfrac{\rule{1cm}{0.4pt}}{\rule{1cm}{0.4pt}} = \dfrac{\rule{1cm}{0.4pt}}{\rule{1cm}{0.4pt}}$

Find the equivalent expression: **(e)** $\dfrac{7x}{18x^2} \bullet \dfrac{\rule{1cm}{0.4pt}}{\rule{1cm}{0.4pt}} = \dfrac{\rule{1cm}{0.4pt}}{\rule{1cm}{0.4pt}}$

Step 3: Add the rational expressions found in Step 2.

Use $\dfrac{a}{c} + \dfrac{b}{c} = \dfrac{a+b}{c}$:

(f) _____

Step 4: Simplify, if possible.

What is the sum, simplified to lowest terms?   **(g)** _____

3. Find the sum and simplify, if possible: $\dfrac{4}{x+2} + \dfrac{2}{x+1}$ *(See textbook Example 3)*   3. _____

4. Find the difference and simplify, if possible: $\dfrac{2}{x} - \dfrac{4}{x-2}$ *(See textbook Example 5)*

Step 1: Find the least common denominator.	Determine the LCD:	**(a)** _____
Step 2: Write each rational expression as an equivalent expression using the LCD.	Find the equivalent expression:	**(b)** $\dfrac{2}{x} \cdot \dfrac{}{} = \dfrac{}{}$
	Find the equivalent expression:	**(c)** $\dfrac{4}{x-2} \cdot \dfrac{}{} = \dfrac{}{}$
Step 3: Subtract the rational expressions found in Step 2.	Use $\dfrac{a}{c} - \dfrac{b}{c} = \dfrac{a-b}{c}$:	**(d)** _____
Step 4: Simplify, if possible.	What is the difference, simplified to lowest terms?	**(e)** _____

5. Find the sum and simplify, if possible: $\dfrac{2y^2+4y-5}{y^2-9} + \dfrac{y^2+4y+4}{9-y^2}$ *(See textbook Example 8)*   5. _____

6. Find the difference and simplify, if possible: $3 - \dfrac{x-2}{x-4}$ *(See textbook Example 9)*   6. _____

Do the Math Exercises 7.5
Adding and Subtracting Rational Expressions with Unlike Denominators

In Problems 1–6, find each sum and simplify.

1. $\dfrac{7}{12}+\dfrac{3}{4}$

2. $\dfrac{5}{2x}+\dfrac{6}{5}$

3. $\dfrac{2}{x-1}+\dfrac{x-1}{x+1}$

4. $\dfrac{7}{n-4}+\dfrac{8}{4-n}$

5. $\dfrac{2x-6}{x^2-x-6}+\dfrac{x+4}{x+2}$

6. $\dfrac{3}{x-4}+\dfrac{x+4}{x^2-16}$

1. _____

2. _____

3. _____

4. _____

5. _____

6. _____

In Problems 7–10, find each difference and simplify, if necessary.

7. $\dfrac{8}{21}-\dfrac{6}{35}$

8. $\dfrac{x-2}{x+2}-\dfrac{x+2}{x-2}$

7. _____

8. _____

9. $\dfrac{x}{x+4} - \dfrac{-4}{x^2+8x+16}$

10. $\dfrac{p+1}{p^2-2p} - \dfrac{2p+3}{2-p}$

9. _____

10. _____

In Problems 11–16, find the LCD of the rational expressions. Then rewrite each as an equivalent rational expression with the LCD.

11. $\dfrac{5}{4n} - \dfrac{3}{n^2}$

12. $\dfrac{x-2}{x+3} + 2$

11. _____

12. _____

13. $\dfrac{-1}{3n-n^2} + \dfrac{1}{3n^2-9n}$

14. $\dfrac{a-5}{2a^2-6a} - \dfrac{6}{12a^2-4a^3}$

13. _____

14. _____

15. $\dfrac{x+1}{x-3} + \dfrac{x+2}{x-2} - \dfrac{x^2+3}{x^2-5x+6}$

16. $\dfrac{7}{w-3} - \dfrac{5}{w} - \dfrac{2w+6}{w^2-9}$

15. _____

16. _____

Name:
Instructor:

Date:
Section:

Five-Minute Warm-Up 7.6
Complex Rational Expressions

1. Factor: $2x^2 - 5x - 12$ 1. _____

2. Factor: $3x^2 - 75$ 2. _____

3. Find the quotient: $\dfrac{x-2}{6} \div \dfrac{x^2-4}{8}$ 3. _____

4. Find the product: $\dfrac{x-1}{x-2} \bullet \left(x^2 - 3x + 2\right)$ 4. _____

5. Find the product: $\left(\dfrac{3}{x} + \dfrac{2}{x^2}\right) \bullet x^3$ 5. _____

Guided Practice 7.6
Complex Rational Expressions

Objective 1: Simplify a Complex Rational Expression by Simplifying the Numerator and Denominator Separately (Method I)

1. In your own words, define a *complex rational expression*.

2. Simplify $\dfrac{\dfrac{5}{x} - \dfrac{x}{5}}{\dfrac{1}{5} - \dfrac{5}{x^2}}$ using Method I. *(See textbook Examples 2 and 3)*

Step 1: Write the numerator of the complex fraction as a single rational expression.	Determine the LCD of x and 5:	**(a)** _____
	Write the equivalent rational expressions on the LCD and then use $\dfrac{a}{c} - \dfrac{b}{c} = \dfrac{a-b}{c}$:	**(b)** _____
Step 2: Write the denominator of the complex fraction as a single rational expression.	Determine the LCD of x^2 and 5:	**(c)** _____
	Write the equivalent rational expressions on the LCD and then add use $\dfrac{a}{c} - \dfrac{b}{c} = \dfrac{a-b}{c}$:	**(d)** _____
Step 3: Rewrite the complex rational expression using the rational expressions determined in Steps 1 and 2.		**(e)** _____
Step 4: Simplify the rational expression using the techniques for dividing rational expressions from textbook Section 7.2.	Rewrite the division problem as a multiplication problem:	**(f)** _____
	Divide out common factors and express the answer as a simplified rational expression in factored form.	**(g)** _____

Objective 2: Simplify a Complex Rational Expression Using the Least Common Denominator (Method II)

3. Method II uses the LCD to simplify complex rational expressions. We use several of the properties of real numbers to simplify the complex rational expression; that is, to find an equivalent rational expression which has a single fraction bar. State the property of real numbers that is illustrated below.

(a) $\dfrac{x-9}{x-9} = 1$

3a. _____

(b) $\dfrac{2}{x} \cdot \dfrac{x-9}{x-9} = \dfrac{2}{x}$

3b. _____

(c) $\dfrac{2}{x} \cdot \dfrac{x-9}{x-9} = \dfrac{2x-18}{x^2-9x}$

3c. _____

4. Simplify $\dfrac{1 + \dfrac{1}{x}}{1 - \dfrac{1}{x^2}}$ using Method II. *(See textbook Examples 5 and 6)*

Step 1: Find the least common denominator among all the denominators in the complex rational expression.

Determine the LCD of x and x^2:

(a) _____

Step 2: Multiply both the numerator and denominator of the complex rational expression by the LCD found in Step 1.

(b) _____

Distribute the LCD to each term:

(c) _____

Step 3: Simplify the rational expression.

Divide out the common factors:

(d) _____

Simplify the expression:

(e) _____

Name:

Instructor:

Date:

Section:

Do the Math Exercises 7.6
Complex Rational Expressions

In Problems 1–4, simplify the complex rational expression using Method I.

1. $\dfrac{\dfrac{4}{t^2}-1}{\dfrac{t+2}{t^3}}$

2. $\dfrac{\dfrac{2}{x}+\dfrac{3}{x^2}}{\dfrac{2x+3}{x}}$

1. _____

2. _____

3. $\dfrac{\dfrac{5}{n-1}+3}{n-\dfrac{2}{n-1}}$

4. $\dfrac{\dfrac{2}{a+b}}{\dfrac{1}{a}+\dfrac{1}{b}}$

3. _____

4. _____

In Problems 5–8, simplify the complex rational expression using Method II.

5. $\dfrac{\dfrac{3c}{4}+\dfrac{3d}{10}}{\dfrac{3c}{2}-\dfrac{6d}{5}}$

6. $\dfrac{\dfrac{a}{2}+4}{2-\dfrac{a}{2}}$

5. _____

6. _____

7. $\dfrac{\dfrac{x}{x+1}}{1+\dfrac{1}{x-1}}$

8. $\dfrac{6x+\dfrac{3}{y}}{\dfrac{9x+3}{y}}$

7. _____

8. _____

In Problems 9–14, simplify the complex rational expression using either Method I or Method II.

9. $\dfrac{1-\dfrac{4}{x^2}}{1-\dfrac{1}{x}-\dfrac{6}{x^2}}$

10. $\dfrac{\dfrac{-6}{y^2+5y+6}}{\dfrac{2}{y+3}-\dfrac{3}{y+2}}$

9. _____

10. _____

11. $\dfrac{2-\dfrac{3}{x}-\dfrac{2}{x^2}}{1-\dfrac{5}{x}+\dfrac{6}{x^2}}$

12. $\dfrac{1-\dfrac{n^2}{25m^2}}{1-\dfrac{n}{5m}}$

11. _____

12. _____

13. $\dfrac{\dfrac{x}{x+y}-1}{\dfrac{y}{x+y}-1}$

14. $\dfrac{\dfrac{5}{2b+3}+\dfrac{1}{2b-3}}{\dfrac{6b}{8b^2-18}}$

13. _____

14. _____

15. **Finding the Mean** The arithmetic mean of a set of numbers is found by adding the numbers and then dividing by the number of entries on the list. Write a complex rational expression to find the arithmetic mean of the expressions $\dfrac{n}{6}, \dfrac{n+3}{2}$, and $\dfrac{2n-1}{8}$ and then simplify the complex rational expression.

15. _____

Five-Minute Warm-Up 7.7
Rational Equations

1. Solve: $-8 + 4(x + 6) = 10x$

1. _____

2. Factor: $x^2 - 3x - 28$

2. _____

3. Solve: $4p^2 - 1 = 0$

3. _____

4. Find the values for which the expression $\dfrac{x - 2}{x^2 + 6x - 16}$ is undefined.

4. _____

5. Solve for P: $A = P + Prt$

5. _____

Guided Practice 7.7
Rational Equations

Objective 1: Solve Equations Containing Rational Expressions

1. When solving rational equations it is important to identify the restrictions on the variable. We exclude all

values of the variable that result in _____.

2. Solve: $\dfrac{x+5}{x-7} = \dfrac{x-3}{x+7}$ *(See textbook Example 2)*

Step 1: Determine the value(s) of the variable that result in an undefined rational expression.	For what values of x will the denominator be equal to zero? **(a)** _____
Step 2: Determine the least common denominator (LCD) of all the denominators.	LCD: **(b)** _____
Step 3: Multiply both sides of the equation by the LCD and simplify the expression on each side of the equation.	Multiply both sides by the LCD: **(c)** _____
	Divide out common factors: **(d)** _____
	Multiply: **(e)** _____
Step 4: Solve the resulting equation.	Solve for x: **(f)** _____
Step 5: Check Verify your solution using the original equation.	Write the solution set: **(g)** _____

3. Solve: $\dfrac{9}{x} + \dfrac{4}{5x} = \dfrac{7}{10}$ *(See textbook Example 3)*

   (a) Determine the excluded values of the variable.   3a. _____

   (b) Determine the LCD of all of the denominators.   3b. _____

   (c) Multiply both sides of the equation by the LCD. What is the resulting equation?   3c. _____

   (d) Solve the resulting equation.   3d. _____

4. Solve: $\dfrac{6}{x^2-1} = \dfrac{5}{x-1} - \dfrac{3}{x+1}$ *(See textbook Example 6)*

    (a) Determine the excluded values of the variable.   4a. _____

    (b) Determine the LCD of all of the denominators.   4b. _____

    (c) Multiply both sides of the equation by the LCD and solve the resulting equation.   4c. _____
    What is the solution to this equation?

    (d) What is the solution set?   4d. _____

Objective 2: Solve for a Variable in a Rational Equation

5. Solve: $\dfrac{6}{z} = \dfrac{2}{x} + \dfrac{1}{y}$ for y *(See textbook Examples 8 and 9)*

Step 1: Determine the value(s) of the variable(s) that result in an undefined rational expression.	For what values of the variables will the denominator **(a)** _____ be equal to zero?
Step 2: Determine the least common denominator (LCD) of all the denominators.	LCD: **(b)** _____
Step 3: Multiply both sides of the equation by the LCD and simplify the expression on each side of the equation.	Multiply both sides by the LCD: **(c)** _____
	Divide out common factors: **(d)** _____
Step 4: Solve the resulting equation for *y*.	Subtract $2yz$ from both sides: **(e)** _____
	Factor out *y*: **(f)** _____
	Divide by the coefficient of *y:* **(g)** _____
	Simplify, if possible. Write the solution: **(h)** _____

Do the Math Exercises 7.7
Rational Equations

In Problems 1–10, solve each equation and state the solution set. Remember to identify the values of the variable for which the expressions in each rational equation are undefined.

1. $\dfrac{4}{p} - \dfrac{5}{4} = \dfrac{5}{2p} + \dfrac{3}{8}$

2. $\dfrac{4}{x-4} = \dfrac{5}{x+4}$

1. _____

2. _____

3. $\dfrac{x-2}{4x} - \dfrac{x+2}{3x} = \dfrac{1}{6x} + \dfrac{2}{3}$

4. $\dfrac{6}{t} - \dfrac{2}{t-1} = \dfrac{2-4t}{t^2-t}$

3. _____

4. _____

5. $\dfrac{4}{x-3} - \dfrac{3}{x-2} = \dfrac{4x+9}{x^2-5x+6}$

6. $\dfrac{2x+3}{x-1} - 2 = \dfrac{3x-1}{4x-4}$

5. _____

6. _____

7. $x = \dfrac{6-5x}{6x}$

8. $\dfrac{2x}{x+4} = \dfrac{x+1}{x+2} - \dfrac{7x+12}{x^2+6x+8}$

7. _____

8. _____

9. $\dfrac{2}{b+2} - \dfrac{5b+6}{b^2-b-6} = \dfrac{-b}{b-3}$

10. $\dfrac{5}{x-2} - \dfrac{2}{2-x} = \dfrac{4}{x+1}$

9. _____

10. _____

In Problems 10–14, solve the equation for the indicated variable.

11. $\dfrac{a}{b+2} = c$ for b

12. $\dfrac{3}{i} - \dfrac{4}{j} = \dfrac{8}{k}$ for j

11. _____

12. _____

13. $B = \dfrac{k}{x} - \dfrac{m}{cx}$ for x

14. $X = \dfrac{ab}{a-b}$ for a

13. _____

14. _____

15. Drug Concentration The concentration C of a drug in a patient's bloodstream in milligrams per liter t hours after ingestion is modeled by $C = \dfrac{40t}{t^2+3}$. When will the concentration of the drug be 10 milligrams per liter?

15. _____

16. Average Cost Suppose that the average daily cost $\overline{C}$ in dollars of manufacturing x bicycles is given by the equation $\overline{C} = \dfrac{x^2+75x+5000}{x}$. Determine the level of production for which the average daily cost will be $240.

16. _____

Five-Minute Warm-Up 7.8
Models Involving Rational Equations

1. Solve: $\dfrac{20}{r} = \dfrac{15}{r-3}$

1. _____

2. Solve: $\dfrac{1}{x} + \dfrac{1}{x-2} = \dfrac{3}{4}$

2. _____

3. Solve: $\dfrac{2}{r} + \dfrac{8}{r+3} = 2$

3. _____

Guided Practice 7.8
Models Involving Rational Equations

Objective 1: Model and Solve Ratio and Proportion Problems

1. Write an example of a ratio.

1. _____

2. Write an example of a proportion.

2. _____

3. Solve: $\dfrac{5}{k+4} = \dfrac{2}{k-1}$ *(See textbook Example 1)*

3. _____

4. Flight Accidents According to the Statistical Abstract of the United States, in 2001, there were 1.22 fatal airplane accidents per 100,000 flight hours. Also, in 2001, there were a total of 321 fatal accidents. How many flight hours were flown in 2001? *(See textbook Example 2)*

4. _____

Objective 2: Model and Solve Problems with Similar Figures

5. In your own words, what does it mean if two geometric figures are *similar*?

6. Suppose that a 6-foot-tall man casts a shadow of 3.2 feet. At the same time of day, a tree casts a shadow of 8 feet. How tall is the tree? *(See textbook Example 5)*

   (a) Write a proportion that can be used to solve this problem.

6a. _____

   (b) Solve the proportion. How tall is the tree?

6b. _____

Objective 3: Model and Solve Work Problems

7. Problems of this type involve completing a job or a task when working at a constant rate. We convert the time t that it takes to complete the job into a unit rate. That is, if it takes t hours to complete a job, then $\dfrac{1}{t}$ of the job is completed per hour.

   (a) If it takes 15 minutes to complete a job, what part of the job is completed per minute? 7a. _____

   (b) If it takes t hours to complete a job, what part of the job is completed per hour?   7b. _____

   (c) If it takes $t + 3$ hours to complete a job, what part of the job is competed per hour?   7c. _____

8. Josh can clean the math building on his campus in 3 hours. Ken takes 5 hours to clean the same building. If they work together, how long will it take for Josh and Ken to clean the math building? *(See textbook Example 6)*

> **Step 1: Identify** We want to know how long it will take Josh and Ken working together to clean the building.
>
> **Step 2: Name** We let *t* represent the time (in hours) that it takes to clean the building when working together.
>
> **Step 3: Translate** What fraction of the job is completed in one hour when working individually and when working together? Write the following ratios:
>
> **(a)** Part of the job completed by Josh in one hour: 8a. _____
>
> **(b)** Part of the job completed by Ken in one hour: 8b. _____
>
> **(c)** Part of the job completed when working together in one hour: 8c. _____
>
> **(d)** Write the model for this problem:
>
> **Step 4: Solve** the equation from Step 3.   8d. _____
>
>
>
>
>
> **Step 5: Check** Is your answer reasonable?
>
> **(e) Step 6: Answer** the question.   8e. _____

Objective 4: Model and Solve Uniform Motion Problems

9. A small plane can travel 1000 miles with the wind in the same time it can go 600 miles against the wind. If the speed of the plane in still air is 180 mph, what is the speed of the wind? *(See textbook Example 8)*

> **(a)** If *w* is the speed of the wind, what is the rate (in mph) when travelling with the wind: 9a. _____
>
> **(b)** If *w* is the speed of the wind, what is the rate when travelling against the wind: 9b. _____
>
> **(c)** Use $t = \dfrac{d}{r}$ to write a rational expression for the time travelled with the wind:
>
>   9c. _____
>
> **(d)** Use $t = \dfrac{d}{r}$ to write a rational expression for the time travelled against the wind:
>
>   9d. _____
>
> **(e)** Write an equation that can be used to solve this problem: 9e. _____
>
> **(f)** Solve your equation and answer the question.   9f. _____

Do the Math Exercises 7.8
Models Involving Rational Equations

In Problems 1–6, solve the proportion.

1. $\dfrac{6}{5} = \dfrac{8}{3x}$

2. $\dfrac{k}{k+3} = \dfrac{6}{15}$

1. _____

2. _____

3. $\dfrac{n+6}{3} = \dfrac{n+4}{5}$

4. $\dfrac{n-2}{4n+7} = \dfrac{-2}{n+1}$

3. _____

4. _____

5. $\dfrac{4}{x^2} = \dfrac{1}{x+3}$

6. $\dfrac{1}{2z-1} = \dfrac{z+4}{z^2+10z+14}$

5. _____

6. _____

In Problems 7 and 8, write an algebraic expression that represents each phrase.

7. If Brian worked for n days and Kellen worked for half as many days, write an algebraic expression that represents the number of days Kellen worked.

7. _____

8. If the rate of the current in a stream is 3 mph and Betsy can paddle her canoe at a rate of r mph in still water, write an algebraic expression that represents Betsy's rate when she paddles downstream.

8. _____

In Problems 9–14, write an equation that could be used to model each of the following. SOLVE THE EQUATION.

9. **Buying Candy** If it costs $3.50 for 2 lb of candy, how much candy can be purchased for $8.75?

9. _____

_____

10. **Comparing Pesos and Pounds** If 50 Mexican pesos are worth approximately 2.5 British pounds and if Sean takes 80 pounds as spending money in Mexico City, how many pesos will he have?

10. _____

_____

11. **Telephone Pole Shadow** A bush that is 4 m tall casts a shadow that is 1.75 m long. How tall is a telephone pole if its shadow is 8.75 m?

11. _____

_____

12. **Pruning Trees** Martin can prune his fruit trees in 4 hours. His neighbor can prune the same trees for him in 7 hours. If they work together on this job, to the nearest tenth of an hour, how long will it take to prune the trees?

12. _____

_____

13. **Travel by Plane** A small plane can travel 1000 miles with the wind in the same time it can go 600 miles against the wind. If the speed of the plane in still air is 180 mph, what is the speed of the wind?

13. _____

_____

14. **A Bike Trip** A bicyclist rides his bicycle 12 miles up a hill and then 16 miles on level terrain. His speed on level ground is 3 miles per hour faster than his speed going uphill. The cyclist rides for the same amount of time going uphill and on level ground. Find his speed going uphill.

14. _____

_____

Five-Minute Warm-Up 7.9
Variation

1. Solve for k:

    (a) $30 = 4k$

    (b) $8 = \dfrac{k}{4}$

    1a. _____

    1b. _____

    (c) $\dfrac{3}{5} = \dfrac{k}{\frac{10}{9}}$

    (d) $31.5 = 5.25k$

    1c. _____

    1d. _____

2. Multiply and round your answer to the nearest hundredth: $4.2 \bullet 3.69$

    2. _____

3. Find the product and simplify: $\dfrac{125}{40} \bullet \dfrac{32}{200}$

    3. _____

Guided Practice 7.9
Variation

Objective 1: Model and Solve Direct Variation Problems

1. We say that y varies directly with x, or y is directly proportional to x, if there is a nonzero number k such

that _____.

2. The number k is called the _____.

3. Write the direct variation equation and then use the given values to solve for the unknown.
(See textbook Examples 1 and 2)

    (a) If y varies directly as x and $y = 6$ when $x = -9$, find an equation that relates
    x and y.   3a. _____

    (b) Suppose that y is directly proportional to x and when $x = -12$, $y = 5$.
    Find y when $x = 20$.   3b. _____

    (c) Suppose the C varies directly as n and when $C = 15.24$, $n = 12$. Find C when
    $n = 37$.   3c. _____

4. **Mortgage Payments** The monthly payment p on a mortgage varies directly with the
    amount borrowed b. Suppose that you decide to borrow \$120,000 using a 15-year
    mortgage at 5.5%. You are told that your payment is \$960.00. (*See textbook Example 2*)

    (a) Write an equation that relates the monthly payment p to the amount borrowed b
    for a mortgage with the same terms.   4a. _____

    (b) Assume that you have decided to buy a more expensive home that requires you to
    borrow \$150,000. What will your monthly payment be?   4b. _____

Objective 2: Model and Solve Inverse Variation Problems

5. We say that y varies inversely with x, or y is inversely proportional to x, if there is a nonzero number k

such that _____.

6. Suppose that y varies inversely with x. When $x = 4$, $y = 12$. Find y when $x = 18$. Write   6. _____
the inverse variation equation, calculate the constant, k, and then use the given values to
solve for the unknown. *(See textbook Example 3)*

7. Electric Current If the voltage of an electric circuit is held constant, the current, I, varies inversely with
the resistance, R. If the current is 60 amperes when the resistance is 150 ohms, find the current when the
resistance is 120 ohms. *(See textbook Example 4)*

  (a) Write the inverse variation equation.

                                                                     7a. _____

  (b) Substitute the given values to find the constant, k.   7b. _____

  (c) Rewrite the inverse variation equation using your constant, k. Substitute the value
for the resistance, R, and solve the equation for the current, I.   7c. _____

Do the Math Exercises 7.9
Variation

In Problems 1 and 2, suppose y varies directly with x. Find an equation that relates x and y given the following.

1. $y = 15$ when $x = 45$ **2.** $y = -10$ when $x = 24$

1. _____

2. _____

In Problems 3 and 4, use the direct variation model to find each of the following.

3. Suppose d varies directly with t.
   (a) Find an equation that relates d and t if it is known that $d = 320$ when $t = 8$.
   (b) Use this equation to determine d if t is 5.

3a. _____

3b. _____

4. Suppose n varies directly with m.
   (a) Find an equation that relates n and m if it is known that $m = 12$ when $n = 9$.
   (b) Use this equation to determine m if n is $-\dfrac{8}{5}$.

4a. _____

4b. _____

In Problems 5 and 6, suppose y varies inversely with x. Find an equation that relates x and y given the following.

5. $x = 7$ when $y = 10$ **6.** $x = \dfrac{3}{4}$ when $y = -\dfrac{2}{9}$

5. _____

6. _____

In Problems 7 and 8, use the inverse variation model to find each of the following.

7. Suppose y varies inversely with x.
   (a) Find an equation that relates y and x if it is known that $y = 4$ when $x = 3$.
   (b) Use this equation to determine y if x is 18.

7a. _____

7b. _____

8. Suppose f varies inversely with d.

    (a) Find an equation that relates f and d if it is known that $f = \dfrac{2}{9}$ when $d = \dfrac{15}{2}$.

    8a. _____

    (b) Use this equation to determine d if f is $\dfrac{15}{4}$.

    8b. _____

In Problems 9 and 10, find the quantity indicated.

9. m varies directly with r. If $r = 24$ when $m = 9$, find r when $m = 24$.

10. x is inversely proportional to y. If $y = 14$ when $x = 4$, find x when $y = \dfrac{2}{3}$.

9. _____

10. _____

11. House of Representatives Apportionment The number of representatives that each state has in the U.S. House of Representatives is directly proportional to the state's population. According to the 2000 Census, California's 53 delegates represent a state whose population is 33.931 million. Find the number of representatives from Florida if its population is 16.029 million.

11. _____

12. Buying Gasoline The cost to purchase a tank of gasoline varies directly with the number of gallons purchased. You notice that the person in front of you spent $61.50 on 15 gallons of gas. If your SUV needs 35 gallons of gas, how much will you spend?

12. _____

13. Measuring Sound The frequency of sound varies inversely with the wavelength. If a radio station broadcasts at a frequency of 90 megahertz, the wavelength of the sound is approximately 3 meters. Find the wavelength of a radio broadcast at 20 megahertz.

13. _____

14. Driving to School The time t that it takes to drive to school varies inversely with your average speed s. It takes you 24 minutes to drive to school when your average speed is 35 miles per hour. Suppose that your average speed to school yesterday was 30 miles per hour. How long did it take you to get to school?

14. _____

Sullivan/Struve/Mazzarella, *Elementary Algebra*, 2e

Five-Minute Warm-Up 8.1
Introduction to Square Roots

In Problems 1 – 6, use the set $\left\{-\dfrac{9}{4}, \sqrt{5}, \dfrac{8}{-2}, 7.2, \pi, \dfrac{0}{15}, 100, \dfrac{-5}{0}\right\}$.

1. Which of the elements of the set are natural numbers?

1. _____

2. Which of the elements of the set are whole numbers?

2. _____

3. Which of the elements of the set are integers?

3. _____

4. Which of the elements of the set are rational numbers?

4. _____

5. Which of the elements of the set are irrational numbers?

5. _____

6. Which of the elements of the set are real numbers?

6. _____

7. Evaluate each of the following:

   (a) $\left(\dfrac{9}{4}\right)^2$ (b) $(0.3)^2$

7a. _____

7b. _____

   (c) $(-6)^2$ (d) -6^2

7c. _____

7d. _____

8. Evaluate: $\left|-\dfrac{3}{5}\right|$

8. _____

Guided Practice 8.1
Introduction to Square Roots

Objective 1: Evaluate Square Roots

1. Fill in the blank: *Properties of Square Roots*

   (a) Every positive real number has two square roots, one _____ and one _____.   1a. _____

   (b) We use the symbol $\sqrt{}$, called a _____, to denote the nonnegative square root.   1b. _____

   (c) The nonnegative square root is called the _____ _____.   1c. _____

   (d) The number under the radical is called the _____.   1d. _____

   (e) $4 = \sqrt{16}$, because $4^2 =$ _____. In general $b = \sqrt{a}$ because _____.   1e. _____

   (f) The square root of a negative real number, such as $\sqrt{-4}$, is   1f. _____

   ____ _____ _____ _____.

2. Find the square roots: (a) 4   (b) 0 *(See textbook Example 1)*   2a. _____

        2b. _____

3. Find the square roots: (a) $\sqrt{\dfrac{49}{64}}$ (b) $\sqrt{0.36}$ *(See textbook Example 2)*   3a. _____

        3b. _____

4. Evaluate: (a) $\sqrt{121}$ (b) $-\sqrt{81}$ *(See textbook Example 3)*   4a. _____

        4b. _____

5. Evaluate: (a) $\sqrt{0.04}$ (b) $-\sqrt{\dfrac{1}{25}}$ *(See textbook Example 4)*   5a. _____

        5b. _____

6. Evaluate: (a) $\sqrt{144 + 25}$ (b) $\sqrt{144} + \sqrt{25}$ *(See textbook Example 6)*   6a. _____

        6b. _____

Objective 2: Determine Whether a Square Root Is Rational, Irrational, or Not a Real Number

7. Fill in the blank:

    (a) The square root of a perfect square is a _____ number.

    7a. _____

    (b) The square root of a positive rational number that is not a perfect square, such as $\sqrt{20}$, is an _____ number.

    7b. _____

    (c) The square root of a negative real number, such as $\sqrt{-2}$ is not a _____ number.

    7c. _____

8. Determine if each square root is rational, irrational, or not a real number. *(See textbook Example 8)*

    8a. _____

    (a) $\sqrt{72}$ **(b)** $\sqrt{-9}$ **(c)** $\sqrt{\dfrac{16}{49}}$

    8b. _____

    8c. _____

Objective 3: Find Square Roots of Variable Expressions

9. Simplify each square root for any real number a and n. *(See textbook Example 9)*

    (a) $\sqrt{a^2}$ **(b)** $\sqrt{(n+9)^2}$

    9a. _____

    9b. _____

Do the Math Exercises 8.1
Introduction to Square Roots

In Problems 1–4, find the value(s) of each expression.

1. the square root of 4

2. the square root of $\dfrac{9}{4}$

1. _____

2. _____

3. the square root of 49

4. the square root of 0.16

3. _____

4. _____

In Problems 5–10, find the exact value of each square root without a calculator.

5. $\sqrt{169}$

6. $\sqrt{0.09}$

5. _____

6. _____

7. $-4\sqrt{4}$

8. $\sqrt{81+144}$

7. _____

8. _____

9. $\sqrt{81}+\sqrt{144}$

10. $\sqrt{100-(4)(8)(3)}$

9. _____

10. _____

Do the Math Exercises 8.1

In Problems 11 and 12, use a calculator or Appendix A to find the approximate value of the square root, rounded to the indicated place.

11. $\sqrt{12}$ to 3 decimal places

12. $\sqrt{18}$ to the nearest hundredth

11. _____

12. _____

In Problems 13–16, tell if the square root is rational, irrational, or not a real number. If the square root is rational, find the exact value; if the square root is irrational, write the approximate value rounded to two decimal places.

13. $\sqrt{-100}$

14. $\sqrt{900}$

13. _____

14. _____

15. $\sqrt{\dfrac{49}{64}}$

16. $\sqrt{24}$

15. _____

16. _____

In Problems 17 and 18, simplify each square root.

17. $\sqrt{(y-16)^2}$, $y-16 \geq 0$

18. $\sqrt{(p+q)^2}$

17. _____

18. _____

Sullivan/Struve/Mazzarella, *Elementary Algebra*, 2e

Five-Minute Warm-Up 8.2
Simplifying Square Roots

1. List the perfect square integers that are less than 150.

_____

2. List the perfect cubed integers that are less than 150.

_____

3. Simplify each of the following expressions, if possible.

(a) $\sqrt{81}$

(b) $\sqrt{\dfrac{16}{9}}$

3a. _____

3b. _____

(c) $\sqrt{-25}$

(d) $-\sqrt{49}$

3c. _____

3d. _____

(e) $\sqrt{0.04}$

(f) $4\sqrt{100}$

3e. _____

3f. _____

Guided Practice 8.2
Simplifying Square Roots

Objective 1: Use the Product Rule to Simplify Square Roots of Constants

1. A square root expression is *simplified* when

2. If $\sqrt{a}$ and $\sqrt{b}$ are nonnegative real numbers, then $\sqrt{ab}$ = _____ • _____ .   2. _____

3. Simplify: $\sqrt{50}$ *(See textbook Example 1)*

Step 1: Write the radicand as the product of factors, one of which is a perfect square.
What perfect square is a factor of 50?   **(a)** _____

Step 2: Write the radicand as the product of two or more radicals, one of which contains the perfect square.
$\sqrt{50}$ =   **(b)** _____

Step 3: Take the square root of any perfect square.   **(c)** _____

4. Simplify each of the following expressions. *(See textbook Examples 2 and 3)*

(a) $\sqrt{24}$ **(b)** $-5\sqrt{27}$ **(c)** $\dfrac{-6+\sqrt{48}}{2}$ 4a. _____

4b. _____

4c. _____

Objective 2: Use the Product Rule to Simplify Square Roots of Variable Expressions

5. Simplify each of the following expressions. Assume each variable represents a nonnegative real number. *(See textbook Examples 4 – 6)*

(a) $\sqrt{64x^{10}}$ **(b)** $\sqrt{45a^2b^6}$ **(c)** $\sqrt{75x^4y^7}$ 5a. _____

5b. _____

5c. _____

Objective 3: Use the Quotient Rule to Simplify Square Roots of Constants and Variable Expressions

6. If $\sqrt{a}$ and $\sqrt{b}$ are nonnegative real numbers and $b \neq 0$, then $\sqrt{\dfrac{a}{b}} = $.

6. _____

7. Simplify each of the following expressions. *(See textbook Examples 7 – 9)*

(a) $\sqrt{\dfrac{5}{9}}$ 　　　　　**(b)** $\sqrt{\dfrac{144a^3}{b^4}}$, $b > 0$ 　　　　**(c)** $\sqrt{\dfrac{27x^{11}}{3x^3}}$, $x > 0$

7a. _____

7b. _____

7c. _____

Do the Math Exercises 8.2
Simplifying Square Roots

In Problems 1–4, simplify each square root.

1. $\sqrt{27}$ **2.** $\sqrt{180}$

1. _____

2. _____

3. $\dfrac{5 - \sqrt{100}}{5}$ **4.** $\dfrac{10 - \sqrt{75}}{5}$

3. _____

4. _____

In Problems 5–10, simplify each square root. Assume all variables represent nonnegative real numbers.

5. $\sqrt{25x^8}$ **6.** $\sqrt{32p^{16}}$

5. _____

6. _____

7. $\sqrt{81y^3}$ **8.** $\sqrt{60n^7}$

7. _____

8. _____

9. $\sqrt{50a^6b^5}$ **10.** $\sqrt{125s^2t^6}$

9. _____

10. _____

In Problems 11–16, simplify each square root. Assume all variables represent positive real numbers.

11. $\sqrt{\dfrac{9}{16}}$

12. $\sqrt{\dfrac{13}{4}}$

11. _____

12. _____

13. $\sqrt{\dfrac{x^{10}}{36}}$

14. $\sqrt{\dfrac{a^7}{b^6}}$

13. _____

14. _____

15. $\sqrt{\dfrac{24a^3b^6}{c^2}}$

16. $\sqrt{\dfrac{36b^{10}}{144b^4}}$

15. _____

16. _____

In Problems 17–20, simplify each expression. Assume all variables represent positive real numbers.

17. $\sqrt{18x^2y^8z^8}$

18. $\sqrt{30m^6np^7}$

17. _____

18. _____

19. $\dfrac{-6+\sqrt{48}}{8}$

20. $\dfrac{-6+\sqrt{108}}{6}$

19. _____

20. _____

Five-Minute Warm-Up 8.3
Adding and Subtracting Square Roots

1. Are the following expressions *like terms*? (Yes or No)

   (a) $4x$ and $8x$ (b) $9x$ and $9y$ (c) x^2 and x 1a. _____

                                                                              1b. _____

                                                                             1c. _____

2. Add: $3x + 8y + 6y + x$ 2. _____

3. Subtract: $\left(a^2 + 2a - 6\right) - \left(a^2 - a + 3\right)$ 3. _____

4. Simplify: $12x\left(3x^2 - x + 1\right)$ 4. _____

5. Factor: $ax + bx$ 5. _____

6. Simplify: $\sqrt{32}$ 6. _____

Guided Practice 8.3
Adding and Subtracting Square Roots

Objective 1: Add and Subtract Square Root Expressions with Like Square Roots

1. Fill in the blank.

    (a) Square root expressions are *like square roots* if each square root has the same _____.

    (b) We you combine like square roots, you apply the _____ Property "in reverse."

1a. _____

1b. _____

2. Perform the indicated operations. Assume each variable represents a nonnegative real number. *(See textbook Examples 1 and 2)*

    (a) $6\sqrt{10} + 2\sqrt{10}$

    (b) $7\sqrt{5} + 2\sqrt{6} - 3\sqrt{5}$

2a. _____

2b. _____

2c. _____

    (c) $\sqrt{x} + 4\sqrt{x} - 9\sqrt{x}$

Objective 2: Add and Subtract Square Root Expressions with Unlike Square Roots

3. Perform the indicated operations. *(See textbook Examples 3 – 5)*

    (a) $\sqrt{8} + \sqrt{18}$

    (b) $\sqrt{45} - \sqrt{125}$

3a. _____

3b. _____

    (c) $-2\sqrt{32} - 3\sqrt{12}$

    (d) $-6b\sqrt{3a^3b} - \left(-2a\sqrt{75ab^3}\right)$

3c. _____

3d. _____

(e) $\sqrt{72x} - \sqrt{18x}$ **(f)** $\sqrt{48x^3} + x\sqrt{75x}$ 3e. _____

3f. _____

(g) $\frac{1}{2}\sqrt{12} - \frac{5}{8}\sqrt{108}$ **(h)** $\sqrt{50} - 2\sqrt{18} + 3\sqrt{20}$ 3g. _____

3h. _____

Do the Math Exercises 8.3
Adding and Subtracting Square Roots

In Problems 1–4, add or subtract the square root expressions.

1. $\sqrt{2} + 7\sqrt{2}$

2. $k\sqrt{11k} + k\sqrt{11k}$

1. _____

2. _____

3. $8\sqrt{3} - 9\sqrt{3}$

4. $-14\sqrt{30} - \left(-\sqrt{30}\right)$

3. _____

4. _____

In Problems 5–12, add or subtract the square root expressions. Assume that variables represent positive real numbers.

5. $\sqrt{8} - \sqrt{32}$

6. $-\sqrt{8} + 3\sqrt{18}$

5. _____

6. _____

7. $-16\sqrt{2ab} + 7\sqrt{18ab}$

8. $2\sqrt{27} + 5\sqrt{48}$

7. _____

8. _____

9. $-3\sqrt{63} + 2\sqrt{28} - 4\sqrt{49}$

10. $\sqrt{80y} - \sqrt{20y}$

9. _____

10. _____

11. $\dfrac{5}{2}\sqrt{300a^5} - a^2\sqrt{27a}$

12. $\dfrac{3}{5}\sqrt{27} - \dfrac{2}{3}\sqrt{12}$

11. _____

12. _____

In Problems 13–18, add or subtract as indicated. Assume all variables represent positive real numbers.

13. $13\sqrt{3p} - 6\sqrt{3p} - 4\sqrt{3}$

14. $3\sqrt{12x^2} - 2\sqrt{27x^2}$

13. _____

14. _____

15. $\sqrt{63xy^2} + \sqrt{28xy^2} - \sqrt{20xy^2}$

16. $3n^4\sqrt{54n^4} - 2n\sqrt{150n^6} + 2\sqrt{24n^9}$

15. _____

16. _____

17. $2 + 3\sqrt{8} - \sqrt{32} - 2\sqrt{18}$

18. $\sqrt{\dfrac{2}{9}} + \sqrt{\dfrac{2}{49}}$

17. _____

18. _____

19. $3\sqrt{\dfrac{3}{4}} - \dfrac{1}{2}\sqrt{\dfrac{3}{16}}$

20. $\dfrac{4}{3}\sqrt{\dfrac{18}{6}} + \dfrac{1}{2}\sqrt{\dfrac{72}{6}}$

19. _____

20. _____

Sullivan/Struve/Mazzarella, *Elementary Algebra*, 2e

Five-Minute Warm-Up 8.4
Multiplying Expressions with Square Roots

1. Find the product: $\left(8n^4\right)^2$

1. _____

2. Find the product: $2x\left(9x^2 + 3x - 2\right)$

2. _____

3. Find the product: $\left(3a - 2b\right)\left(a + 3b\right)$

3. _____

4. Find the product: $\left(2x - 5\right)^2$

4. _____

5. Find the product: $\left(3x + 7y\right)\left(3x - 7y\right)$

5. _____

6. Simplify each expression.

(a) $\sqrt{75x^3}$, $x \geq 0$ **(b)** $3\sqrt{20} - 4\sqrt{125}$

6a. _____

6b. _____

Guided Practice 8.4
Multiplying Expressions with Square Roots

Objective 1: Find the Product of Square Roots Containing One Term

1. The Product Rule of Square Roots states: if $\sqrt{a}$ and $\sqrt{b}$ are real numbers, then $\sqrt{a} \bullet \sqrt{b} = \sqrt{ab}$.
Use this rule to simplify each of the following. Assume each variable represents a nonnegative real number. *(See textbook Example 1)*

(a) $\sqrt{7} \bullet \sqrt{3}$

(b) $\sqrt{5n} \bullet \sqrt{11m}$

1a. _____

1b. _____

2. Find the product and simplify, if possible. *(See textbook Example 2)*

(a) $\sqrt{8} \bullet \sqrt{2}$

(b) $\sqrt{12} \bullet \sqrt{54}$

2a. _____

2b. _____

3. Multiply and simplify, if possible. Assume each variable represents a nonnegative real number. *(See textbook Examples 3 and 4)*

(a) $\sqrt{5m^5} \bullet \sqrt{15m}$

(b) $\sqrt{3p^2} \bullet \sqrt{18p^3}$

3a. _____

3b. _____

4. Simplify each expression. *(See textbook Examples 5 and 6)*

(a) $\left(-\sqrt{5}\right)^2$

(b) $\left(-3\sqrt{6}\right)\left(4\sqrt{8}\right)$

4a. _____

4b. _____

Objective 2: Find the Product of Square Roots Using the Distributive Property

5. Multiply and simplify, if possible. *(See textbook Example 7)*

(a) $5\left(7 - \sqrt{5}\right)$

(b) $\sqrt{3}\left(7 + 4\sqrt{6}\right)$

5a. _____

5b. _____

Objective 3: Find the Product of Square Roots Using FOIL

6. Find each product and simplify, if possible. *(See textbook Examples 8 and 9)*

(a) $\left(4 - \sqrt{5}\right)\left(6 + \sqrt{5}\right)$

(b) $\left(5 - 2\sqrt{3}\right)\left(2 - 8\sqrt{3}\right)$

6a. _____

6b. _____

Objective 4: *Find the Product of Square Roots Using Special Products*

7. We can use special products to simplify products containing square roots. These rules are:

(a) $(A + B)^2 = $ _____ **(b)** $(A - B)^2 = $ _____

(c) $(A + B)(A - B) = $ _____

8. Find each product. *(See textbook Examples 10 and 11)*

(a) $\left(3 - 2\sqrt{5}\right)^2$ **(b)** $\left(5 - 2\sqrt{10}\right)\left(5 + 2\sqrt{10}\right)$ 8a. _____

8b. _____

Do the Math Exercises 8.4
Multiplying Expressions with Square Roots

In Problems 1–8, find the product and simplify, if possible. Assume all variables represent nonnegative real numbers.

1. $-\sqrt{10} \cdot \sqrt{15}$

2. $\sqrt{90} \cdot \sqrt{54}$

1. _____

2. _____

3. $-\sqrt{2a^2} \cdot \sqrt{50a^6}$

4. $\sqrt{6q^3} \cdot \sqrt{8q^4}$

3. _____

4. _____

5. $8\sqrt{15} \cdot 3\sqrt{20}$

6. $10\sqrt{6n^2} \cdot 5\sqrt{18n}$

5. _____

6. _____

7. $\left(-\sqrt{42}\right)^2$

8. $\left(\sqrt{6y}\right)^2$

7. _____

8. _____

In Problems 9 and 10, find the product and simplify. Assume all variables represent nonnegative real numbers.

9. $\sqrt{8}\left(2 + 3\sqrt{8}\right)$

10. $\sqrt{5}\left(5 + 3\sqrt{15}\right)$

9. _____

10. _____

In Problems 11–14, find the product using the FOIL method, and simplify. Assume all variables represent nonnegative real numbers.

11. $\left(8-\sqrt{2}\right)\left(7-\sqrt{2}\right)$

12. $\left(4+\sqrt{11}\right)\left(2-\sqrt{11}\right)$

11. _____

12. _____

13. $\left(7+3\sqrt{3}\right)\left(2-3\sqrt{2}\right)$

14. $\left(7\sqrt{x}-3\sqrt{2x}\right)\left(2\sqrt{x}+3\sqrt{2x}\right)$

13. _____

14. _____

In Problems 15–18, find the product using special products: $(A + B)^2$, $(A - B)^2$, and $(A - B)(A + B)$, and simplify. Assume all variables represent nonnegative real numbers.

15. $\left(3-\sqrt{2}\right)^2$

16. $\left(2\sqrt{x}+3\right)^2$

15. _____

16. _____

17. $\left(4-3\sqrt{5}\right)\left(4+3\sqrt{5}\right)$

18. $\left(5\sqrt{3}-\sqrt{7}\right)\left(5\sqrt{3}+\sqrt{7}\right)$

17. _____

18. _____

Five-Minute Warm-Up 8.5
Dividing Expressions with Square Roots

1. Find the quotient: $\dfrac{24x^4 y}{10x^3 y^3}$

   1. _____

2. Find the quotient: $\sqrt{\dfrac{36x^3}{y^8}}$, $x \geq 0$, $y \geq 0$

   2. _____

3. What would be needed to multiply 12 by in order to make the smallest perfect square that is a multiple of 12?

   3. _____

4. Multiply: $\dfrac{\sqrt{6}}{\sqrt{3}} \cdot \dfrac{\sqrt{8}}{\sqrt{3}}$

   4. _____

5. Multiply: $\left(4 - \sqrt{2}\right)\left(4 + \sqrt{2}\right)$

   5. _____

Guided Practice 8.5
Dividing Expressions with Square Roots

Objective 1: Find the Quotient of Two Square Roots

1. Simplify each expression. *(See textbook Example 1)*

   (a) $\dfrac{\sqrt{6}}{\sqrt{24}}$

   (b) $\dfrac{\sqrt{14a}}{\sqrt{50a}}$

   1a. _____

   1b. _____

Objective 2: Rationalize a Denominator Containing One Term

2. In your own words, what does it mean to rationalize the denominator of a rational expression?

3. To rationalize a denominator containing a single square root, we multiply the numerator and denominator

of the quotient by a square root so that the radicand in the denominator becomes _____.

4. Determine what to multiply each quotient by so that the denominator contains a radicand which is a perfect square. *(See textbook Example 3)*

   (a) $\dfrac{3}{\sqrt{3}}$

   (b) $\dfrac{4}{\sqrt{20}}$

   (c) $\dfrac{1}{\sqrt{8a}}$

   4a. _____

   4b. _____

   4c. _____

5. Rationalize the denominator of the expression $\dfrac{2}{\sqrt{3}}$. *(See textbook Example 3)*

Step 1: Determine the number that you multiply the radicand in the denominator by to obtain a perfect square.	What number do you need to multiply the denominator by so that it is a perfect square?	**(a)** _____
	Write an expression for one, written as the number from (a) over itself:	**(b)** _____
Step 2: Multiply the numerators; multiply the denominators.	Multiply $\dfrac{2}{\sqrt{3}}$ by the expression in (b):	**(c)** _____
Step 3: Simplify, if possible.	Simplify, if possible. What expression is equivalent to $\dfrac{2}{\sqrt{3}}$ with a rational denominator?	**(d)** _____

6. Rationalize the denominator of each expression. *(See textbook Example 4)*

(a) $\dfrac{9}{4\sqrt{6}}$

(b) $\sqrt{\dfrac{5}{27}}$

6a. _____

6b. _____

Objective 3: *Rationalize a Denominator Containing Two Terms*

7. To rationalize a denominator containing two terms, we multiply both numerator and denominator by the

_____ of the denominator.

8. Identify the conjugate of each expression. Then multiply the expression by its conjugate. *(See textbook Example 5)*

(a) $2 - \sqrt{5}$

(b) $3\sqrt{3} + 5\sqrt{2}$

8a. _____

8b. _____

9. Rationalize the denominator of the expression $\dfrac{8}{2 + \sqrt{6}}$. *(See textbook Example 6)*

Step 1: Multiply the numerator and denominator of the quotient by the conjugate of the denominator.	Identify the conjugate of the denominator: **(a)** _____
	Write an expression for one, written as the number from (a) over itself: **(b)** _____
Step 2: Multiply the numerators; multiply the denominators.	Multiply $\dfrac{8}{2 + \sqrt{6}}$ by the expression in (b). Leave the numerator in factored form: **(c)** _____
Step 3: Simplify, if possible.	Divide out the common factor and then distribute. What expression is equivalent to $\dfrac{8}{2 + \sqrt{6}}$ with a rational denominator? **(d)** _____

10. Rationalize the denominator: $\dfrac{\sqrt{xy}}{\sqrt{x} - \sqrt{y}}$ *(See textbook Example 7)*

10. _____

Do the Math Exercises 8.5
Dividing Expressions with Square Roots

In Problems 1–4, find the product and simplify, if possible. Assume all variables represent nonnegative real numbers.

1. $-\dfrac{\sqrt{10}}{\sqrt{90}}$

2. $\dfrac{\sqrt{144}}{\sqrt{3}}$

1. _____

2. _____

3. $\dfrac{-\sqrt{18x^3}}{\sqrt{2x}}$

4. $\sqrt{\dfrac{147a^3b^4}{3ab^4}}$

3. _____

4. _____

In Problems 5–8, rationalize the denominator of each expression. Assume all variables represent positive real numbers.

5. $\dfrac{9}{\sqrt{11}}$

6. $-\dfrac{15}{2\sqrt{3}}$

5. _____

6. _____

7. $\dfrac{4p}{\sqrt{12p^5}}$

8. $-\dfrac{5\sqrt{a}}{2\sqrt{10b}}$

7. _____

8. _____

In Problems 9 and 10, determine the conjugate of each expression. Then find the product of the expression and its conjugate.

9. $\sqrt{10}+3$

10. $\sqrt{x}-2\sqrt{y}$

9. _____

10. _____

In Problems 11–14, rationalize the denominator of each expression. Assume all variables represent positive real numbers.

11. $\dfrac{4}{4-\sqrt{10}}$

12. $\dfrac{8}{\sqrt{x}-2}$

11. _____

12. _____

13. $\dfrac{9}{-2\sqrt{3}+15}$

14. $\dfrac{\sqrt{y}}{1+2\sqrt{y}}$

13. _____

14. _____

In Problems 15 and 16, simplify the expression. Assume all variables represent positive real numbers.

15. $\dfrac{\sqrt{27}+\sqrt{24}}{\sqrt{3}}$

16. $\sqrt{\dfrac{3x}{y}}\cdot\sqrt{\dfrac{y^2}{2x^4}}$

15. _____

16. _____

Five-Minute Warm-Up 8.6
Solving Equations Containing Square Roots

1. Solve: $8x + 4 = 0$

1. _____

2. Solve: $2x^3 - 14x^2 + 24x = 0$

2. _____

3. Find the product: $(4x + 1)^2$

3. _____

4. Find the product: $(3\sqrt{2} - 4)^2$

4. _____

5. Find the product: $(\sqrt{x + 4})^2$, $x \geq -4$

5. _____

6. Solve: $x^2 + x = 6$

6. _____

Guided Practice 8.6
Solving Equations Containing Square Roots

Objective 1: Determine Whether or Not a Number is a Solution of a Radical Equation

1. Determine if the number is a solution to the equation. *(See textbook Example 1)*

   (a) $\sqrt{12x + 45} = 3$; $x = -3$ **(b)** $\sqrt{6x + 16} = -2$; $x = -2$

1a. _____

1b. _____

Objective 2: Solve Equations Containing One Square Root

2. Solve: $\sqrt{3x - 5} = 4$ *(See textbook Example 2)*

Step 1: Isolate the radical.	This is already done.	$\sqrt{3x - 5} = 4$
Step 2: Square both sides of the equation.	Square both sides:	**(a)** _____
Step 3: Solve the equation that results.	Add 5 to both sides:	**(b)** _____
	Divide both sides by 3:	**(c)** _____
Step 4: Check	Verify that your result satisfies the original equation. Be careful! There are extraneous solutions when working with radical equations.	
	State the solution set:	**(d)** _____

3. Solve: $\sqrt{2x + 1} - 2 = 3$ *(See textbook Example 3)*

Step 1: Isolate the radical.	Add 2 to both sides:	**(a)** _____
Step 2: Square both sides of the equation.	Square both sides.	**(b)** _____
Step 3: Solve the equation that results.	Subtract 1 from both sides:	**(c)** _____
	Divide both sides by 2:	**(d)** _____
Step 4: Check	Verify that your result satisfies the original equation. Be careful! There are extraneous solutions when working with radical equations.	
	State the solution set:	**(e)** _____

Objective 3: Solve Equations that Involve Two Square Roots

4. Solve: $\sqrt{3x+1} = \sqrt{2x-7}$ *(See textbook Examples 7 and 8)*

4. _____

5. Solve: $\sqrt{2x-1} - \sqrt{x-1} = 1$ *(See textbook Example 9)*

(a) Since this equation has two radical expressions, we want one radical on the left side of the equation and the other radical expression on the right side of the equation. Write this equation so that one radical is on each side of the equation.

5a. _____

(b) Square both sides of the equation:

5b. _____

(c) Use $(A+B)^2 = A^2 + 2AB + B^2$ to square $\left(\sqrt{x-1}+1\right)$:

5c. _____

(d) Use $\left(\sqrt{A+B}\right)^2 = A + B$ to square $\left(\sqrt{2x-1}\right)^2$:

5d. _____

(e) Isolate the remaining radical. Since the coefficient of the radical expression does not divide evenly into the other side of the equation, we will leave the coefficient of the radical.

5e. _____

(f) Use $(A-B)^2 = A^2 - 2AB + B^2$, $(ab)^2 = a^2b^2$ and $\left(\sqrt{A-B}\right)^2 = A - B$ to square both sides:

5f. _____

(g) What equation is left to solve?

5g. _____

(h) Solve the resulting equation. What value(s) did you find for x?

5h. _____

(i) Check and state the solution set:

5i. _____

Sullivan/Struve/Mazzarella, *Elementary Algebra*, 2e

Do the Math Exercises 8.6
Solving Equations Containing Square Roots

In Problems 1–4, determine if the number is a solution to the equation.

1. $\sqrt{4n-7} = 3;\ n = 4$

2. $\sqrt{3y-4} + 5 = 0;\ y = -7$

1. _____

2. _____

3. $\sqrt{p+6} = p+4;\ p = -5$

4. $\sqrt{3y+15} = \sqrt{y+6} + 1;\ y = -2$

3. _____

4. _____

In Problems 5–10, solve the equation and check the solution.

5. $\sqrt{3-y} = 2$

6. $10 = 23 + \sqrt{p}$

5. _____

6. _____

7. $y = \sqrt{24-5y}$

8. $\sqrt{2x-8} = x-4$

7. _____

8. _____

9. $\sqrt{17x+13} - 5 = x$

10. $4 + \sqrt{2x} - x = 0$

9. _____

10. _____

In Problems 11–14, solve the equation and check the solution.

11. $\sqrt{5x+1} = \sqrt{3x-15}$

12. $\sqrt{2x^2+8} - \sqrt{x^2-3x+6} = 0$

11. _____

12. _____

13. $\sqrt{x+3} = \sqrt{x+15}$

14. $2\sqrt{s-4} = 2 + \sqrt{4s}$

13. _____

14. _____

In Problems 15 and 16, solve the equation and check the solution.

15. $2x-1 = \sqrt{4x^2-11}$

16. $1 = \sqrt{3-3c} - 2c$

15. _____

16. _____

17. Braking Distance Dale is driving his Hummer down a county highway made of asphalt in the desert. If he is going 60 mph and slams on the brakes, how far will the car skid before it comes to a stop? Use $S = \sqrt{15d}$, if d is the distance in feet to stop a car going S miles per hour.

17. _____

18. Fun with Numbers The square root of 5 less than twice a number is 7. Find the number.

18. _____

Five-Minute Warm-Up 8.7
Higher Roots and Rational Exponents

1. Evaluate each of the following expressions.

   (a) 3^4　　　　　(b) $(-2)^4$　　　　　(c) -4^4

   1a. _____

   1b. _____

   1c. _____

2. Simplify: $5x^3 \cdot 2x^6$

   2. _____

3. Simplify: $\dfrac{18x^3 y^2}{24xy^5}$

   3. _____

4. Simplify: $\left(-5x^4\right)^3$

   4. _____

5. Simplify: $\dfrac{n^4 \cdot n^{-9}}{n^{-2}}$

   5. _____

6. Simplify: $\dfrac{\sqrt{75}}{\sqrt{3}}$

   6. _____

Guided Practice 8.7
Higher Roots and Rational Exponents

Objective 1: Evaluate Higher Roots

1. In the notation $\sqrt[3]{8} = 2$, 3 is called the _____, 8 is called the _____, and 2 is the _____.

2. If the index is even, then the radicand must _____ in order for the radical to simplify to real number.

If the index is odd, then the radicand can be _____ and the expression will simplify to any real number.

3. Since $(-4)^2 = 16$ and $4^2 = 16$, it could be interpreted that $\sqrt{16} = -4$ or 4. In fact, $\sqrt{16} = 4$ only, because when the index even, we use the _____ _____, which must be ≥ 0.

4. Evaluate each root without using a calculator. *(See textbook Example 1)*

   (a) $\sqrt[3]{-27}$ (b) $\sqrt{-36}$ (c) $\sqrt[4]{\dfrac{16}{81}}$ (d) $\sqrt[5]{32}$ 4a. _____

   4b. _____

   4c. _____

   4d. _____

Objective 2: Use Product and Quotient Rules to Simplify Higher Roots
Assume each variable represents a nonnegative real number.

5. Simplify each of the following expressions. *(See textbook Example 3)*

   (a) $\sqrt[3]{16}$ (b) $\sqrt[3]{81n^5}$ (c) $\sqrt[3]{-54p^6}$ 5a. _____

   5b. _____

   5c. _____

6. Simplify each of the following expressions. *(See textbook Example 4)*

   (a) $\dfrac{\sqrt[3]{-216}}{\sqrt[3]{3}}$ (b) $\dfrac{\sqrt[4]{1440z^7}}{\sqrt[4]{2z^3}}$ 6a. _____

   6b. _____

Objective 3: Define and Evaluate Expressions of the Form $a^{\frac{1}{n}}$

7. Write each expression as a radical and evaluate, if possible. *(See textbook Example 5)*

   (a) $144^{\frac{1}{2}}$ (b) $(-64)^{\frac{1}{2}}$ (c) $-36^{\frac{1}{2}}$ 7a. _____

   7b. _____

   7c. _____

Objective 4: Define and Evaluate Expressions of the Form $a^{\frac{m}{n}}$

8. If a is a real number and $\dfrac{m}{n}$ is a rational number in lowest terms with $n \geq 2$, then

$$a^{\frac{m}{n}} = \underline{\hspace{1.5in}} \quad \text{or} \quad \underline{\hspace{1in}}, \text{ provided that } \sqrt[n]{a} \text{ exists.}$$

9. Evaluate each of the following expressions, if possible. *(See textbook Examples 8 and 9)*

   (a) $36^{\frac{3}{2}}$ **(b)** $-9^{\frac{5}{2}}$ **(c)** $(-64)^{\frac{2}{3}}$ **(d)** $(-81)^{\frac{3}{2}}$

9a. _____

9b. _____

9c. _____

9d. _____

10. Write each radical with a rational exponent. *(See textbook Example 10)*

   (a) $\sqrt[4]{x^5}$ **(b)** $\left(\sqrt[3]{2x^2}\right)^4$

10a. _____

10b. _____

11. Write each exponential expression as a radical. *(See textbook Example 11)*

   (a) $3x^{\frac{2}{3}}$ **(b)** $(3x)^{\frac{2}{3}}$

11a. _____

11b. _____

12. Rewrite each of the following with a positive exponent and evaluate, if possible. *(See textbook Example 12)*

   (a) $8^{-\frac{2}{3}}$ **(b)** $-16^{-\frac{3}{4}}$

12a. _____

12b. _____

Objective 5: Use Laws of Exponents to Simplify Expressions with Rational Exponents

13. Simplify each of the following expressions. *(See textbook Example 13)*

   (a) $6^{\frac{12}{5}} \cdot 6^{-\frac{2}{5}}$ **(b)** $\left(49x^{\frac{2}{3}}\right)^{\frac{1}{2}}$ **(c)** $\dfrac{\left(-8a^4\right)^{\frac{1}{3}}}{a^{\frac{5}{6}}}$

13a. _____

13b. _____

13c. _____

Do the Math Exercises 8.7
Higher Roots and Rational Exponents

In Problems 1–10, evaluate or simplify each expression.

1. the cube root of 27

2. $\sqrt[4]{-81}$

1. _____

2. _____

3. $\sqrt[4]{81}$

4. $\sqrt[5]{32}$

3. _____

4. _____

5. $\sqrt[12]{z^{12}}, z \geq 0$

6. $\sqrt[3]{24}$

5. _____

6. _____

7. $\dfrac{\sqrt[4]{64}}{\sqrt[4]{4}}$

8. $-\sqrt[3]{-250}$

7. _____

8. _____

9. $\dfrac{\sqrt[3]{320w^7}}{\sqrt[3]{5w}}$

10. $\dfrac{\sqrt[4]{96a^{13}}}{\sqrt[4]{2a}}$

9. _____

10. _____

In Problems 11–14, write each expression as a radical and evaluate or simplify.

11. $(125p)^{\frac{1}{3}}$

12. $(-16)^{\frac{5}{4}}$

11. _____

12. _____

13. $-16^{-\frac{3}{2}}$

14. $(4x)^{\frac{3}{4}}$

13. _____

14. _____

In Problems 15 and 16, write each radical as an expression with a rational exponent.

15. $-4\sqrt[3]{z^2}$

16. $\sqrt[4]{(7y)^3}$

15. _____

16. _____

In Problems 17–20, use Laws of Exponents to simplify each of the following. All variables represent positive real numbers.

17. $6^{\frac{12}{5}} \bullet 6^{-\frac{2}{5}}$

18. $(16y^{\frac{1}{3}})^{\frac{3}{4}}$

17. _____

18. _____

19. $\dfrac{x^{-\frac{3}{5}} \bullet x^{\frac{9}{5}}}{x^{\frac{2}{5}}}$

20. $\dfrac{(3x)^{\frac{5}{4}}}{(3x)^{-\frac{3}{4}}}$

19. _____

20. _____

Sullivan/Struve/Mazzarella, *Elementary Algebra, 2e*

Five-Minute Warm-Up 9.1
Solving Quadratic Equations Using the Square Root Property

1. Simplify each of the following.

   (a) $\sqrt{169}$ (b) $-2\sqrt{16}$ 1a. _____

   1b. _____

2. Simplify each of the following.

   (a) $\sqrt{20}$ (b) $\sqrt{75}$ 2a. _____

   2b. _____

3. Factor: $4z^2 - 12z + 9$ 3. _____

4. Given the irrational number $\sqrt{32}$,

   (a) Express the radical as an exact value.   (b) Approximate the radical to the nearest
                                               hundredth.   4a. _____

   4b. _____

5. Simplify: $\dfrac{12 - \sqrt{128}}{4}$ 5. _____

6. Rationalize the denominator: $\dfrac{12}{\sqrt{3}}$ 6. _____

Guided Practice 9.1
Solving Quadratic Equations Using the Square Root Property

Objective 1: Solve Quadratic Equations Using the Square Root Property

1. State the Square Root Property: If $x^2 = p$, where x is any variable expression and p is a real number,

then _____.

2. If the solution to a quadratic equation is $x = -1 \pm 2\sqrt{3}$, write the solution set. _____

3. *True or False:* $\sqrt{x^2 - 16} = \sqrt{81}$ simplifies to $x - 4 = \pm 9$. _____

4. Solve: $n^2 - 45 = 0$ *(See textbook Example 1)*

Step 1: Isolate the expression containing the squared term.		$n^2 - 45 = 0$
	Add 45 to both sides:	**(a)** _____
Step 2: Use the Square Root Property. Don't forget the $\pm$ symbol.	Take the square root of both sides of the equation:	**(b)** _____
	Simplify the radical:	**(c)** _____
Step 3: Isolate the variable, if necessary. **Step 4:** Verify your solution(s).		The variable is already isolated.
	State the solution set:	**(d)** _____

5. Solve the quadratic equation: $5x^2 - 12 = 0$ *(See textbook Example 3)*   5. _____

6. Solve: $(8x-4)^2 - 3 = 17$ *(See textbook Example 4)*

Step 1: Isolate the expression containing the squared term.

$(8x-4)^2 - 3 = 17$

Add 3 to both sides:

(a) _____

Step 2: Use the Square Root Property. Don't forget the $\pm$ symbol.

Take the square root of both sides of the equation:

(b) _____

Simplify the radical:

(c) _____

Step 3: Isolate the variable, if necessary.

Add 4 to both sides:

(d) _____

Divide both sides by 8:

(e) _____

Factor the numerator:

(f) _____

Divide out common factors:

(g) _____

Step 4: Verify your solution(s).

State the solution set:

(h) _____

7. Solve: $x^2 - 10x + 25 = 72$ *(See textbook Example 5)*

7. _____

Objective 2: Solve Problems Using the Pythagorean Theorem

8. State the Pythagorean Theorem in words. If x and y are the lengths of the legs and z is the length of the hypotenuse, write an equation which uses these variables to state the Pythagorean Theorem.

9. In a right triangle, one leg is of length 8 inches. If the other leg has length of 20 inches,

(a) what is the exact length of the hypotenuse? *(See textbook Example 7)*

9a. _____

(b) Find the approximate length of the hypotenuse to the nearest tenth of an inch.

9b. _____

Sullivan/Struve/Mazzarella, *Elementary Algebra*, 2e

Do the Math Exercises 9.1
Solving Quadratic Equations Using the Square Root Property

In Problems 1–10, solve each quadratic equation using the Square Root Property. Express radicals in simplest form.

1. $x^2 = 81$

2. $48 = t^2$

1. _____

2. _____

3. $\dfrac{1}{3}w^2 = 9$

4. $64x^2 = 4$

3. _____

4. _____

5. $7x^2 + 8 = 24$

6. $16 = (x+5)^2$

5. _____

6. _____

7. $\dfrac{(z+4)^2}{5} = 3$

8. $(x+3)^2 - 12 = 24$

7. _____

8. _____

9. $(8g-3)^2 - 2 = -1$

10. $(3x-6)^2 - 40 = -48$

9. _____

10. _____

Do the Math Exercises 9.1

In Problems 11 and 12, solve each quadratic equation by first factoring the perfect square trinomial on one side of the equation and then using the Square Root Property.

11. $x^2 + 16x + 64 = 36$ **12.** $16 = 9z^2 - 12z + 4$ 11. _____

12. _____

In Problems 13–16, solve each quadratic equation by either factoring or the Square Root Property. For each problem, choose the method that is most efficient.

13. $x^2 - 14x + 48 = 0$ **14.** $81 = 2x^2 + 9$ 13. _____

14. _____

15. $x^2 + 14x + 49 = 0$ **16.** $\sqrt{p^2 + 4} = 4$ 15. _____

16. _____

17. Picture Frame A rectangular picture frame measures 17 inches on the diagonal and is 8   17. _____
inches high. Exactly how wide is this frame?

       Sullivan/Struve/Mazzarella, *Elementary Algebra*, 2e

Five-Minute Warm-Up 9.2
Solving Quadratic Equations by Completing the Square

1. Factor: $x^2 + 12x + 36$

1. _____

2. Solve: $(2x - 3)^2 = 25$

2. _____

3. Find the quotient: $\dfrac{6x^2 - 12x + 8}{6}$

3. _____

4. Solve: $(x - 3)(x + 2) = 14$

4. _____

5. Simplify: $\left(\sqrt{x - 9}\right)^2$; $x \geq 9$

5. _____

Guided Practice 9.2
Solving Quadratic Equations by Completing the Square

Objective 1: Complete the Square in One Variable

1. If a polynomial is of the form $x^2 + bx + c$, c must be equal to _____ in order to be a perfect square trinomial.

2. Determine the number that must be added to the expression to make it a perfect square trinomial. Then factor the expression. *(See textbook Example 1)*

(a) $p^2 - 14p$ **(b)** $n^2 + 9$ **(c)** $z^2 + \dfrac{4}{3}z$ 2a. _____

2b. _____

2c. _____

Objective 2: Solve Quadratic Equations by Completing the Square

3. Solve: $p^2 - 6p - 18 = 0$ *(See textbook Example 2)*

Step 1: Rewrite $x^2 + bx + c = 0$ as $x^2 + bx = -c$ by adding or subtracting the constant from both sides of the equation.	Add 18 to both sides:	$p^2 - 6p - 18 = 0$ **(a)** _____
Step 2: Complete the square on the expression $x^2 + bx$. Remember, whatever is added to one side of the equation must also be added to the other side!	What value must be added to both sides to make the expression on the left a perfect square trinomial?	**(b)** _____
	Add this number to both sides of the equation and simplify:	**(c)** _____
Step 3: Factor the perfect square trinomial on the left side of the equation.	Use: $A^2 - 2AB + B^2 = \left(A - B\right)^2$	**(d)** _____
Step 4: Solve the equation using the Square Root Property.	Take the square root of both sides of the equation:	**(e)** _____
	Simplify the square root:	**(f)** _____
	Add 3 to both sides:	**(g)** _____
Step 5: Verify your solutions(s).	State the solution set:	**(h)** _____

4. Solve: $(x+5)(x-2) = -3$ *(See textbook Example 4)*

4.

5. To solve the equation $3x^2 - 9x + 12 = 0$ by the completing the square, the first step is_____.

6. Solve: $x(x-4) + 20 = 6$ *(See textbook Example 6)*

6.

Do the Math Exercises 9.2
Solving Quadratic Equations by Completing the Square

In Problems 1–6, determine the number that must be added to the expression to make it a perfect square trinomial. Then factor the expression.

1. $x^2 + 24x$

2. $p^2 - 36p$

1. _____

2. _____

3. $y^2 + \dfrac{6}{7}y$

4. $z^2 - 5z$

3. _____

4. _____

5. $w^2 - w$

6. $r^2 + \dfrac{r}{4}$

5. _____

6. _____

In Problems 7–14, solve the quadratic equation using completing the square.

7. $x^2 + 6x = 40$

8. $z^2 - 5z + 6 = 0$

7. _____

8. _____

9. $x^2 - 8x + 18 = 0$

10. $n^2 = 6n + 18$

9. _____

10. _____

Do the Math Exercises 9.2

11. $2x^2 - 6x + 4 = 0$ **12.** $3x^2 = 4 - 2x$ 11. _____

12. _____

13. $4n^2 + 12n + 1 = 0$ **14.** $-9z = 2z^2 + 44$ 13. _____

14. _____

In Problems 15–20, solve the quadratic equation using factoring, the Square Root Property, or completing the square.

15. $(x - 8)(x - 1) = 1$ **16.** $4x^2 + 12x + 9 = 0$ 15. _____

16. _____

17. $2y^2 + 15 = 51$ **18.** $n^2 - 13n = 30$ 17. _____

18. _____

19. $z^2 - 48 = 0$ **20.** $t(2t + 9) = -4$ 19. _____

20. _____

Sullivan/Struve/Mazzarella, *Elementary Algebra*, 2e

Five-Minute Warm-Up 9.3
Solving Quadratic Equations Using the Quadratic Formula

1. Evaluate $3 \pm \sqrt{9 - 4(2)(1)}$. Give the exact value.

1. _____

2. Evaluate $\dfrac{-4 \pm \sqrt{(-4)^2 - 4(2)(-1)}}{2}$. Give the exact value.

2. _____

3. Write each quadratic equation in standard form. If $ax^2 + bx + c = 0$, identify the coefficients, a, b, and c.

   (a) $15 - p^2 = 3p$ (b) $(n - 2)(2n + 3) = -1$

3a. _____

_____

3b. _____

_____

4. Solve: $\dfrac{7}{2x} + 2 = \dfrac{3}{x}$

4. _____

Guided Practice 9.3
Solving Quadratic Equations Using the Quadratic Formula

Objective 1: Solve Quadratic Equations Using the Quadratic Formula

1. If $ax^2 + bx + c = 0$, then $x =$ _____.

2. When using the quadratic formula, the first step is to write the quadratic equation in _____.

3. Write the quadratic equation in standard form and then identify the values assigned to a, b, and c.
      Do not solve the equation.

    (a) $x - 2x^2 = -4$ **(b)** $2 - 3w^2 = 8$ **(c)** $3y^2 = 6y$ 3a. _____

                                                                       3b. _____

                                                                       3c. _____

4. Solve: $8n^2 - 2n - 3 = 0$ *(See textbook Example 1)*

Step 1: Write the equation in standard form $ax^2 + bx + c = 0$ and identify the values of a, b, and c.

Since the equation is already in standard form, identify the values for a, b, and c.

$$8n^2 - 2n - 3 = 0$$

  (a) $a =$ _____; $b =$ _____; $c =$ _____

Step 2: Substitute the values of a, b, and c into the quadratic formula.

Write the quadratic formula: **(b)** _____

Substitute the values for a, b, and c.   **(c)** _____

Step 3: Simplify the expression found in Step 2.

What is the value of the radicand?   **(d)** _____

Simplify the expression in (c): **(e)** _____

Write the two expressions using $a \pm b$ means $a - b$ or $a + b$:   **(f)** _____

Step 4: Check

State the solution set: **(g)** _____

5. Solve: $x^2 - 6x = 3$ *(See textbook Example 3)*

5. _____

6. Solve: $4x + \dfrac{3}{x} = 2$ *(See textbook Example 4)*

6. _____

Objective 2: Use the Discriminant to Determine Which Method to Use When Solving a Quadratic Equation

7. In the quadratic equation $ax^2 + bx + c = 0$, the *discriminant* is used to determine the nature and number of solutions. It can also be used to help determine which method to use when solving a quadratic equation. To find the discriminant, substitute the identified values for a, b, and c into part of the quadratic

formula, _____.

8. Which method would you use to solve the quadratic equation?

    (a) $b^2 - 4ac$ is a perfect square

8a. _____

    (b) $b^2 - 4ac = 0$

8b. _____

    (c) $b^2 - 4ac$ is positive, but not a perfect square

8c. _____

    (d) $b^2 - 4ac$ is negative

8d. _____

9. Determine the discriminant of each quadratic equation. Use the discriminant to determine the most efficient way of solving the quadratic equation. *Do not solve the equation.* *(See textbook Examples 8 and 9)*

    (a) $4x^2 + 5x - 9 = 0$ **(b)** $x^2 + 4x + 9 = 0$ **(c)** $4x^2 = 4x - 1$

9a. _____

9b. _____

9c. _____

Do the Math Exercises 9.3
Solving Quadratic Equations Using the Quadratic Formula

In Problems 1–8, solve each equation using the quadratic formula.

1. $x^2 - 4x - 6 = 0$

2. $2x^2 + 9x - 5 = 0$

1. _____

2. _____

3. $5y^2 + \dfrac{1}{2}y - 1 = 0$

4. $x^2 = 3x + 9$

3. _____

4. _____

5. $x^2 + 9x = -14$

6. $9x^2 - 12x + 6 = 2$

5. _____

6. _____

7. $\dfrac{x^2}{6} - \dfrac{4}{3}x + \dfrac{8}{3} = 0$

8. $2(t - t^2) = 3$

7. _____

8. _____

In Problems 9–14, determine the discriminant and then use this value to determine the most efficient way of solving the quadratic equation. DO NOT SOLVE THE EQUATION.

9. $3s^2 - s + 4 = 0$

10. $0 = 3x^2 - 8x - 3$

9. _____

10. _____

11. $2x = 1 + x^2$

12. $n(2 + n) = 7$

11. _____

12. _____

13. $0 = 4z - 1 - 4z^2$

14. $\dfrac{x^2}{2} - \dfrac{x}{2} + \dfrac{1}{5} = 0$

13. _____

14. _____

15. Heavenly Spa The monthly revenue from selling x passes to Heavenly Spa is given by the equation $R = 0.02x^2 + 40x$.
(a) What is the revenue during a month that the spa sells 25 passes?
(b) How many passes does Heavenly Spa need to sell to have a monthly income of $2050?

15. _____

Five-Minute Warm-Up 9.4
Problem Solving Using Quadratic Equations

1. Use the Pythagorean Theorem to find b when $a = 4$ and $c = 8$.

1. _____

2. Find the value approximated to the nearest hundredth: $\dfrac{4 \pm \sqrt{27}}{3}$

2. _____

3. Solve: $n^2 - 4n = -2$

3. _____

Guided Practice 9.4
Problem Solving Using Quadratic Equations

Objective 1: Model and Solve Direct Translation Problems

1. The area of a rectangle is 117 square meters. The length of the rectangle is 4 meters more than the width. What are the dimensions of the rectangle? *(See textbook Example 1)*

(a) **Step 1: Identify** To find the dimensions of the rectangle, we need to know the length and width of the rectangle.

    Step 2: Name Let w represent the width of the rectangle. Write an expression for the length of the rectangle.　　1a. _____

(b) **Step 3: Translate** In general, what formula do we use to calculate the area of a rectangle?　　1b. _____

(c) Write an equation that will model the area of this rectangle.　　1c. _____

(d) **Step 4: Solve** Solve your equation in (c). What value(s) did you find for w?　　1d. _____

(e) Since w represents the width of a rectangle, which solution must be discarded?　　1e. _____

    Step 5: Check

(f) **Step 6: Answer the Question**　　1f. _____

Objective 2: Solve Problems Modeled by Quadratic Equations

2. Advertising Costs Joel has started a catering business for which he buys advertising. He predicts that as time goes by, this expense will decrease because he will have all the business he can handle. A model that predicts the monthly advertising expenditures N (in dollars), x years after he opens his business is given by the equation $N = \dfrac{2000}{x^2 + 5}$. How much did Joel spend on advertising during the third year his catering business was in operation? *(See textbook Example 4)*

  (a) Step 1: Identify Here we want to know the value N when $x = ?$. _____

     Step 2: Name The variables are named in the problem. N is monthly advertising expenditures and x is years of operation.

  (b) Step 3: Translate Use the information from (a) and the given formula to write a model for this problem. _____

  (c) Step 4: Solve Simplify your expression in (b). What value did you find for N? _____

  (d) Since N represents the monthly cost, what was Joel's annual expenditure? _____

     Step 5: Check

  (e) Step 6: Answer the Question _____

  (f) After how many years will Joel be spending $25 per month on advertising? _____

3. Punkin Chunkin Suppose that a catapult in the Punkin Chunkin contest releases a pumpkin 8 feet above the ground at an angle of $45°$ to the horizontal with an initial speed 220 feet per second. The model $s(t) = -16t^2 + 155t + 8$ can be used to estimate the height s of an object after t seconds. Round your answers to the nearest tenth. *(See textbook Example 5)*

  (a) After how many seconds will the pumpkin be 40 feet above the ground?   3a. _____

  (b) After how long will the pumpkin strike the ground?   3b. _____

Do the Math Exercises 9.4
Problem Solving Using Quadratic Equations

In Problems 1–2, find the dimensions of each rectangle.

1. The length of a rectangle is five feet less than twice the width. If the area of the rectangle is 150 square feet, find the dimensions.

   1. _____

2. The length of a rectangle is seven centimeters more than half of the width. If the area of the rectangle is 120 square centimeters, find the dimensions.

   2. _____

In Problems 3–4, find the unknown values in each triangle.

3. In a triangle, the height is one inch less than twice the base. If the area of the triangle is 60 square inches, find the base and height of the triangle.

   3. _____

4. In a triangle, the base is seven meters less than the height. If the area of the triangle is 99 square meters, find the base and height of the triangle.

   4. _____

In Problems 5–6, use the Pythagorean Theorem to find the unknown values.

5. In a right triangle, one leg is five kilometers less than the other. If the length of the hypotenuse is 10 kilometers, find the exact lengths of each leg. Then approximate these lengths to the nearest tenth of a kilometer.

   5. _____

6. In a right triangle, the hypotenuse is one yard less than twice one of the legs. If the length of the other leg is 3 yards, find the exact lengths of the remaining leg and hypotenuse. Then approximate these lengths to the nearest tenth of a yard.

   6. _____

In Problems 7–10, write an equation to represent the unknown number and then solve.

7. The product of two consecutive integers is 182. Find two pairs of integers that have this product.

7. _____

8. Consider two consecutive integers. The product of the larger and half of the smaller is 55. Find the positive integers that satisfy these conditions.

8. _____

9. Consider any real number such that its square decreased by four times the number is 14. Find the exact value of the number(s).

9. _____

10. Consider any real number such that the square of the number decreased by four times the number is negative one. Find the exact value of the number(s).

10. _____

11. **Projectile Motion** Omar shot a projectile up vertically with an initial speed of 92 feet per second. If he was standing on a building 25 feet high, the equation that models the height of the projectile is $h = -16t^2 + 92t + 25$.

    (a) How high was the projectile $\dfrac{3}{4}$ of a second after it was launched?

    11a._____

    (b) How high is it after three seconds?

    11b._____

    (c) To the nearest hundredth of a second, how long does it take for the projectile to reach an altitude of 75 feet?

    11c._____

    (d) To the nearest hundredth of a second, how will it take to hit the ground? (*Hint:* It is on the ground when $h = 0$.

    11d._____

12. **Complementary Angles** Two angles are complementary if the sum of their measures is $90°$.

    (a) If the measure of one angle is $y°$, what is the measure of its compliment?

    12a._____

    (b) Half of the product of the measure of an angle and its complement is $700°$. Find the measure of the angle.

    12b._____

Five-Minute Warm-Up 9.5
The Complex Number System

In Problems 1 – 6, use the set $\left\{ \sqrt{-4}, -\dfrac{10}{2}, \sqrt{9}, \dfrac{8}{9}, 0.\overline{3}, \pi, \dfrac{0}{15}, 25, \dfrac{2}{0}, \sqrt{20} \right\}$.

1. Which of the elements of the set are natural numbers?   1. _____

2. Which of the elements of the set are whole numbers?   2. _____

3. Which of the elements of the set are integers?   3. _____

4. Which of the elements of the set are rational numbers?   4. _____

5. Which of the elements of the set are irrational numbers?   5. _____

6. Which of the elements of the set are real numbers?   6. _____

7. Distribute: $\dfrac{7}{3}x^2\left(6x^3 + 9x - 3\right)$ 7. _____

8. Multiply: $(2x + 5)(3x - 1)$ 8. _____

9. Multiply: $\left(5 - \sqrt{2}\right)\left(5 + \sqrt{2}\right)$ 9. _____

Guided Practice 9.5
The Complex Number System

1. The *imaginary unit*, denoted by i, is the number whose square is -1. $i^2 =$ _____ or $i =$ _____.

2. *Complex numbers* are numbers of the form _____ where a and b are _____. When a

number is in a form such as $6 - 2i$, we say that the number is in _____ form. The *real part*

of the complex number is _____ and the *imaginary part* is _____.

Objective 1: Evaluate the Square Root of Negative Real Numbers
3. Write each of the following as a pure imaginary number. *(See textbook Example 1)*

   (a) $\sqrt{-81}$ (b) $-2\sqrt{-11}$ (c) $\sqrt{-27}$

3a. _____

3b. _____

3c. _____

4. Write each of the following in standard form. *(See textbook Examples 2 and 3)*

   (a) $6 + \sqrt{-4}$ (b) $\dfrac{3}{2} - \sqrt{-48}$ (c) $\dfrac{9 - \sqrt{-45}}{-3}$

4a. _____

4b. _____

4c. _____

Objective 2: Add or Subtract Complex Numbers
5. Before beginning any operations with complex numbers, you must write the number

in _____ form.

6. Perform the indicated operations. *(See textbook Examples 4 and 5)*

   (a) $(-5 - 2i) + (4 - 9i)$ (b) $\left(-4 - \sqrt{-24}\right) - \left(-2 + \sqrt{-54}\right)$

6a. _____

6b. _____

Objective 3: Multiply Complex Numbers
7. Multiply each of the following complex numbers. *(See textbook Examples 6 – 9)*

   (a) $\sqrt{-4} \cdot \sqrt{-9}$ (b) $(3 - 5i)^2$

7a. _____

7b. _____

   (c) $\dfrac{3}{2}i(6 - 8i)$ (d) $\left(-2 + \sqrt{-4}\right)\left(-3 - \sqrt{-25}\right)$

7c. _____

Objective 4: Divide Complex Numbers

8. Divide: $\dfrac{20}{3+i}$. *(See textbook Example 11)*

Step 1: Write the numerator and denominator in standard form, $a + bi$.

The numerator and denominator are already in standard form.

Step 2: Multiply the numerator and denominator by the complex conjugate of the denominator.

Identify the conjugate of the denominator:

(a) _____

Multiply the quotient by 1, written with the conjugate:

(b) _____

Step 3: Simplify by writing the quotient in standard form, $a + bi$.

Multiply the numerator; the denominator is of the form $(a + bi)(a - bi) = a^2 + b^2$:

(c) _____

Divide out common factors:

(d) _____

Distribute:

(e) _____

9. Divide: $\dfrac{8 + 26i}{\sqrt{-16}}$ *(See textbook Example 12)*

9. _____

Objective 5: Solve Quadratic Equations with Complex Solutions

10. Solve the quadratic equation: $5x^2 + 2x + 1 = 0$ *(See textbook Example 15)*

10. _____

Do the Math Exercises 9.5
The Complex Number System

In Problems 1–2, write each of the following in standard form.

1. $15 - \sqrt{-45}$

2. $\dfrac{-6 + \sqrt{-12}}{2}$

1._____

2._____

In Problems 3–6, add or subtract as indicated. Express your answer in standard form.

3. $(-5 - 2i) + (4 - 9i)$

4. $(-5 + 4i) - (7 + 4i)$

3._____

4._____

5. $(8 + \sqrt{-25}) + (-3 - \sqrt{-16})$

6. $(13 - \sqrt{-121}) - (8 + \sqrt{-49})$

5._____

6._____

In Problems 7–10, multiply. Express your answer in standard form.

7. $5i(-2 + 7i)$

8. $(-3 - 3i)(1 - 2i)$

7._____

8._____

9. $(-8 + 6i)^2$

10. $(-2 + \sqrt{-4})(-3 - \sqrt{-25})$

9._____

10._____

In Problems 11–12, write the complex conjugate of the given number and then find the product of the complex number and its conjugate.

11. $4 - 5i$ **12.** $2 - \sqrt{-12}$ 11._____

12._____

In Problems 13–16, divide. Express your answer in standard form.

13. $\dfrac{15}{1 - 2i}$ **14.** $\dfrac{2 - 5i}{i}$ 13._____

14._____

15. $\dfrac{-5 - 4i}{5i}$ **16.** $\dfrac{-5}{-3 + i}$ 15._____

16._____

In Problems 17–20, solve the quadratic equation by any method.

17. $2x^2 - 3x + 6 = 0$ **18.** $(z + 6)^2 + 13 = 4$ 17._____

18._____

19. $2x^2 - 90 = 0$ **20.** $x(x + 6) = 9$ 19._____

20._____

Five-Minute Warm-Up 10.1
Quadratic Equations in Two Variables

1. Graph: $y = -\dfrac{2}{3}x + 1$

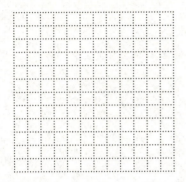

2. Graph: $x = \dfrac{3}{2}$

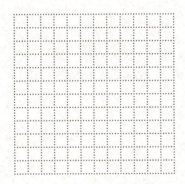

3. Solve: $x^2 + 2x - 48 = 0$ 3. _____

4. Solve: $5x^2 - 2x - 1 = 0$. Give the exact answer and then the result to the nearest tenth.   4. _____

5. Find the intercepts: $9x - 6y = -18$ 5. _____

Guided Practice 10.1
Quadratic Equations in Two Variables

Objective 1: Graph a Quadratic Equation by Plotting Points

1. Graph $y = x^2 - x - 4$ by plotting points. *(See textbook Example 1)*

We begin by creating a table of values.

	x	y	(x, y)
(a)	-2	$y = (\quad)^2 - (\quad) - 4 = $ _____	_____
(b)	-1	$y = (\quad)^2 - (\quad) - 4 = $ _____	_____
(c)	0	$y = (\quad)^2 - (\quad) - 4 = $ _____	_____
(d)	1	$y = (\quad)^2 - (\quad) - 4 = $ _____	_____
(e)	2	$y = (\quad)^2 - (\quad) - 4 = $ _____	_____

(f) Plot the ordered pairs in the last column and connect the points in a smooth curve.

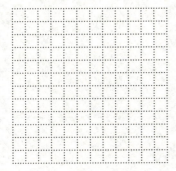

Objective 2: Find the Intercepts of the Graph of a Quadratic Equation

2. The x-intercepts can be found by _____.

3. The y-intercept can be found by _____.

4. Find the intercepts of the graph of $y = -x^2 + 3x + 18$. *(See textbook Example 3)*   4. _____

Objective 3: Find the Vertex and Axis of Symmetry of the Graph of a Quadratic Equation

5. The graph of the quadratic equation $y = ax^2 + bx + c$ opens _____ if a is positive.   5. _____

6. The graph of the quadratic equation $y = ax^2 + bx + c$ opens _____ if a is negative.   6. _____

7. The graph of the quadratic equation $y = ax^2 + bx + c$, $a \neq 0$ has a vertex whose x-coordinate is _____ .   7. _____

Objective 4: Graph a Quadratic Equation Using Its Vertex, Intercepts, and Axis of Symmetry

8. Graph $y = -x^2 + 2x + 8$ using its properties. *(See textbook Example 6)*

Step 1: Determine whether the parabola opens up or down.

Determine the values using $y = ax^2 + bx + c$:

(a) $a =$ _____; $b =$ _____; $c =$ _____

Does the parabola open up or down?

(b) _____

Step 2: Determine the vertex and axis of symmetry.

Calculate the x-coordinate of the vertex $\left(x = -\dfrac{b}{2a} \right)$:

(c) _____

Calculate the y-coordinate of the vertex:

(d) _____

Identify the vertex:

(e) _____

Identify the axis of symmetry:

(f) _____

Step 3: Determine the y-intercept.

Let $x = 0$ and solve for y:

(g) _____

Step 4: Find the discriminant, $b^2 - 4ac$, to determine the number of the x-intercepts. Then determine the x-intercepts, if any.

Evaluate the discriminant:

(h) _____

Number of x-intercepts:

(i) _____

Factor $0 = -x^2 + 2x + 4$ and use the Zero-Product Property to identify the x-intercepts:

(j) _____

Step 5: Plot the points and draw the graph of the quadratic equation. Use the axis of symmetry to find an additional point.

Graph:

(k)

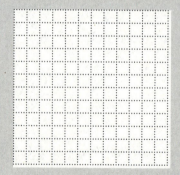

Do the Math Exercises 10.1
Quadratic Equations in Two Variables

Graph the quadratic equation by filling in the table and plotting the points.

1. $y = x^2 - 2$

x	y
-3	
-2	
-1	
0	
1	
2	
3	

In Problems 2–3, determine the x- and y-intercepts of the graph of each quadratic equation. If an intercept is an irrational number, give both the exact and approximate value.

2. $y = -3x^2 - 5x - 2$ **3.** $y = x^2 - 2$

2. _____

3. _____

In Problems 4–5, for each quadratic equation, (a) determine whether the parabola opens up or down, (b) find the vertex of the parabola, and (c) find the axis of symmetry.

4. $y = -x^2 + 3$ **5.** $y = 2x^2 - 9x + 4$

4a. _____

4b. _____

4c. _____

5a. _____

5b. _____

5c. _____

In Problems 6–7, graph each quadratic equation by determining whether the parabola opens up or down, finding the vertex, the intercepts, and the axis of symmetry.

6. $y = -x^2 + 25$

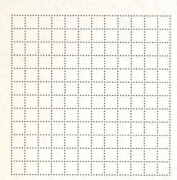

7. $y = 4x^2 + 12x + 9$

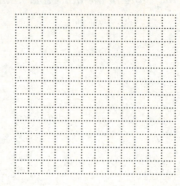

In Problems 8–9, determine whether the quadratic equation has a maximum or minimum value, then find the maximum or minimum value of the equation.

8. $y = 4x^2 - 16x + 19$

9. $y = -2x^2 - 4x + 5$

8. _____

9. _____

10. Height of a Shot Put A track and field athlete throws a shot put, and during one competition, the path of the shot put could be modeled by the quadratic equation

$$h = -0.04x^2 + 3x + 6$$

where h is the height (in feet) of the shot put at the instant it has traveled x feet horizontally.

(a) Determine the distance x at which the height of the shot put is a maximum by finding $x = -\dfrac{b}{2a}$.

10a. _____

(b) Determine the maximum height of the shot put by evaluating the equation at the value of x found in part **(a)**.

10b. _____

Five-Minute Warm-Up 10.2
Relations

1. Plot the points in the rectangular coordinate system.

   $A(-3, 2)$; $B(0, -2)$; $C(4, -1)$; $D(2, 0)$; $E(-1, -3)$

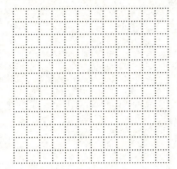

2. Graph: $y = -x$

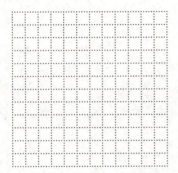

3. Graph: $4x - 2y = -8$

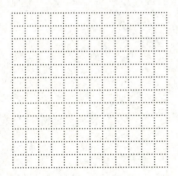

4. Graph: $y = -2x^2 + 3$

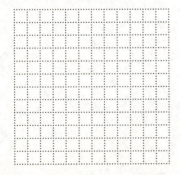

Guided Practice 10.2
Relations

Objective 1: Define Relations

1. In your own words, write a definition for a *relation*.

2. Represent the relation shown in the figure below as a set of ordered pairs. Then state the domain and the range of the relation. *(See textbook Examples 1 and 2)*

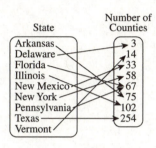

_____

_____

Objective 2: Find the Domain and the Range of a Relation

3. The *domain* is the set of all _____ and is the set of _____ -coordinates for the relation which is defined by the set of ordered pairs (x, y).

4. The *range* is the set of all _____ and is the set of _____ -coordinates for the relation which is defined by the set of ordered pairs (x, y).

*In Problems 5 and 6, the figure shows the graph of a relation. Determine (**a**) the domain and (**b**) the range of the relation. (See textbook Example 4)*

5.

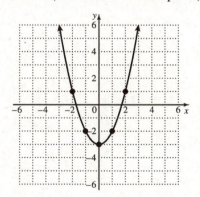

6.

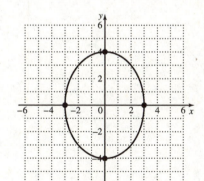

5a. _____

5b. _____

6a. _____

6b. _____

Objective 3: *Graph a Relation Defined by an Equation*

7. Graph the relation $y = x^2 - 3$. Use the graph to determine **(f)** the domain and **(g)** the range of the relation. *(See textbook Example 5)*

	x	y	(x, y)
(a)	-2		
(b)	-1		
(c)	0		
(d)	1		
(e)	2		

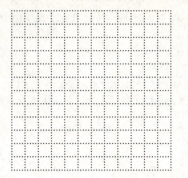

7f. _____

7g. _____

Do the Math Exercises 10.2
Relations

In Problems 1–2, represent each relation as a set of ordered pairs. Then state the domain and range of each relation.

1.

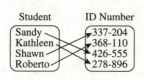

2.

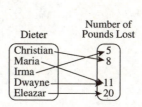

1. _____

2. _____

In Problems 3–4, use the set of ordered pairs to represent the relation as a map. Then state the domain and range of each relation.

3. {(–2, 3); (–4, 3); (–2, 2); (1, 1)}

4. {(3, 0); (–1, –6); (–1, 6); (1, 0)}

3. _____

4. _____

In Problems 5–8, identify the domain and range of the relation shown in the figure.

5.

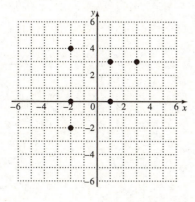

6.

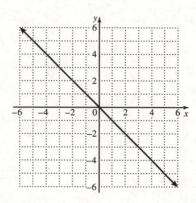

5. _____

6. _____

7.

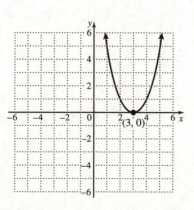

8.

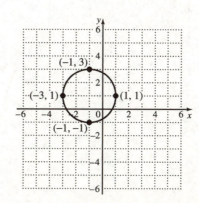

7. _____

8. _____

In Problems 9–14, graph each relation. Use the graph to identify the domain and range of the relation.

9. $y = x + 4$

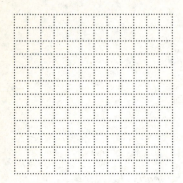

10. $y = 3x - 2$

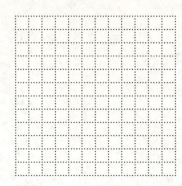

9. _____

_____

10. _____

_____

11. $y = -x^2 + 4x - 4$

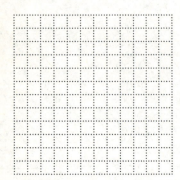

12. $y = 3x^2 + 6x$

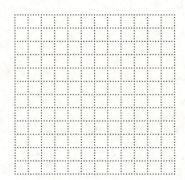

11. _____

_____

12. _____

_____

13. $y = -x^2 + 5x + 2$

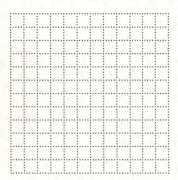

14. $y = x^2 - 5x + 1$

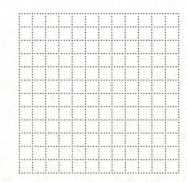

13. _____

_____

14. _____

_____

Sullivan/Struve/Mazzarella, *Elementary Algebra*, 2e

Five-Minute Warm-Up 10.3
An Introduction to Functions

1. Evaluate the expression $2x^2 - 4x + 5$ for

   (a) $x = -4$ (b) $x = 2$ 1a. _____

                                                                 1b. _____

2. Evaluate the expression $-\dfrac{2}{3}x + 2$ for

   (a) $x = 9$ (b) $x = -\dfrac{5}{2}$ 2a. _____

                                                                 2b. _____

3. Evaluate the expression $\left| 3x - 10 \right|$ for $x = -\dfrac{4}{9}$.   3. _____

4. Evaluate the expression $\sqrt{9 - x^2}$ for $x = -1$.   4. _____

5. State (a) the domain and (b) the range: $\{(-2, 4), (-1, 2), (2, 4), (3, 6)\}$

                                                                 5a. _____

                                                                 5b. _____

Guided Practice 10.3
An Introduction to Functions

Objective 1: Determine Whether or Not a Relation Expressed as a Map or Ordered Pairs Represents a Function

1. In your own words, write a definition for a *function*.

2. Determine whether the relation represents a function. If the relation is a function, then state its domain and range. *(See textbook Example 2)*

(a) $\{(-1, -1), (4, 4), (5, 5)\}$ _____

(b) $\{(2, -1), (3, -2), (2, -3)\}$ _____

Objective 2: Determine Whether or Not a Relation Expressed as an Equation Represents a Function

3. The symbol $\pm$ is a shorthand device and is read "plus or minus." Write the two equations that are represented by $y = \pm 2x$ and then determine whether the equation shows y is a function of x.

(a) $y = \pm 2x$ means $y =$ _____ and also $y =$ _____ . (b) Is y a function of x? _____

4. Determine whether each equation shows y as a function of x. *(See textbook Examples 3 and 4)*

(a) $y = 9$ (b) $x = -1$ 4a. _____

4b. _____

(c) $y = \pm \sqrt{x - 2}$ (d) $4x - 3y = -24$ 4c. _____

4d. _____

Objective 3: Identify the Graph of a Function Using the Vertical Line Test

5. We use the Vertical Line Test to determine whether the graph of a relation is a function. In your own words, state the *Vertical Line Test (VLT)*.

6. Which of the graphs are graphs of functions? *(See textbook Example 5)*

6. _____

(a)

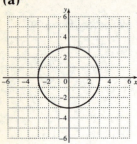

(b)

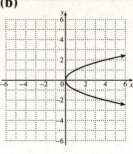

(c)

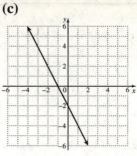

(d)

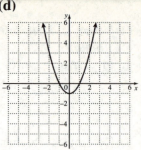

Objective 4: Find the Value of a Function

In Problems 7 – 10, find the value of each function. (See textbook Examples 6 and 7)

7. $f(x) = -2x^2 + 3x$

(a) $f(-4)$ **(b)** $f\left(\dfrac{3}{2}\right)$

8. $g(x) = \dfrac{3}{2}x + 2$

(a) $g(8)$ **(b)** $g(-2)$

7a. _____

7b. _____

8a. _____

8b. _____

9. $h(t) = 4$

(a) $h(0)$ **(b)** $h(3)$

10. $F(z) = z^2 + 4$

(a) $F(-2)$ **(b)** $F(-5)$

9a. _____

9b. _____

10a. _____

10b. _____

Do the Math Exercises 10.3
An Introduction to Functions

In Problems 1–2, determine whether each relation represents a function. If the relation is a function, state its domain and range.

1.

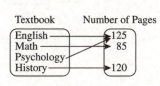

2.

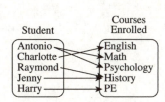

1. _____

2. _____

In Problems 3–4, determine whether each relation represents a function. If the relation is a function, state its domain and range.

3. $\{(3, 1); (2, 1); (0, 1); (-1, 1)\}$ **4.** $\{(5, 0); (3, 1); (-1, 2); (3, 2)\}$

3. _____

4. _____

In Problems 5–8, determine whether each equation represents y as a function of x.

5. $x = -4$ **6.** $y = 0$

5. _____

6. _____

7. $y = x^2 - 3x + 2$ **8.** $y = \sqrt{x}$

7. _____

8. _____

In Problems 9–10, determine whether the graph is that of a function.

9.

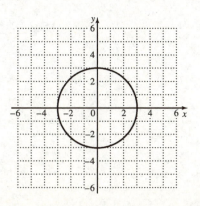

10.

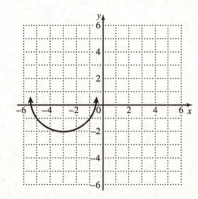

9. _____

10. _____

In Problems 11–15, find the following values for each function:

(a) $f(0)$ **(b)** $f(3)$ **(c)** $f(-2)$

11. $f(x) = 2x + 4$ **12.** $f(x) = 1 - x$

                                          11a. _____

                                          11b. _____

                                          11c. _____

12a. _____

12b. _____

12c. _____

13. $f(x) = -1$ **14.** $f(x) = x^2 + 4$

13a. _____

13b. _____

13c. _____

14a. _____

14b. _____

14c. _____

15. $f(x) = -2x^2 + x - 3$

15a. _____

15b. _____

15c. _____

Chapter 1 Answers

Section 1.2

Five-Minute Warm-Up **1.** Counting numbers such as 1, 2, 3... **2.** Numbers whose only factors are 1 and itself. **3.** Natural numbers which are not prime. 1 is neither prime nor composite. **4.** The LCM of two or more numbers is the smallest number that is a multiple of each of the numbers. **5a.** 4, 9 **5b.** 36 **6.** 24, 48, 72 **7.** 1, 2, 4 **8.** 1, 2, 3, 4, 6, 18 **9.** $2 \bullet 3 \bullet 3$

Guided Practice **1a.** $2 \bullet 3 \bullet 5$ **1b.** $2 \bullet 2 \bullet 2 \bullet 3 \bullet 5$ **2a.** $2 \bullet 2 \bullet 2$ **2b.** $2 \bullet 2 \bullet 3$ **2c.** $2 \bullet 2$ **2d.** $2 \bullet 3$

2e. 24 **3a.** $\dfrac{3}{3}$ **3b.** $\dfrac{15}{24}$ **4a.** $2 \bullet 2 \bullet 3 \bullet 3$ **4b.** $2 \bullet 2 \bullet 2 \bullet 3$ **4c.** $2 \bullet 2 \bullet 2 \bullet 3 \bullet 3$ **4d.** 72 **4e.** $\dfrac{2}{2}$ **4f.** $\dfrac{3}{3}$

4g. $\dfrac{7}{36} = \dfrac{14}{72}; \dfrac{11}{24} = \dfrac{33}{72}$ **5a.** 94,000 **5b.** 94,200 **5c.** 94,205 **5d.** 94,204.6 **5e.** 94,200 **5f.** 94,204.64 **6.** 0.75 **7.** 0.5%

Do the Math **1.** $2 \bullet 3 \bullet 3 \bullet 3$ **2.** $3 \bullet 3 \bullet 7$ **3.** 280 **4.** 180 **5.** $\dfrac{12}{15}$ **6.** $\dfrac{10}{28}$ **7.** $\dfrac{3}{36}$ and $\dfrac{10}{36}$ **8.** $\dfrac{25}{60}$ and $\dfrac{28}{60}$

9. $\dfrac{3}{5}$ **10.** $\dfrac{8}{9}$ **11.** tenths place **12.** ones place **13.** 7300 **14.** 37.4 **15.** $0.\overline{2}$ **16.** 0.34375 **17.** $\dfrac{2}{5}$ **18.** $\dfrac{179}{500}$

19. 0.59 **20.** 34.9%

Section 1.3

Five-Minute Warm-Up **1.** 0 **2.** undefined **3.** 1 **4.** 3 **5a.** 3 **5b.** 7 **6.** $\dfrac{3}{6}$, 0.001, $\dfrac{0}{1}$, $\dfrac{4}{2}$ **7.** $\dfrac{3}{6}$, 0.001, $\dfrac{4}{2}$

8. $\dfrac{3}{6}$, 0.001 **9.** $\dfrac{0}{1}$ **10.** $0.8\overline{3}$ **11.** 0.8

Guided Practice **1a.** $B = \{0, 2, 4, 6, 8\}$ **1b.** $C = \{1, 3, 5, 7, 9\}$ **2a.** $\{14\}$ **2b.** $\left\{14, \dfrac{0}{3}\right\}$

2c. $\left\{14, \dfrac{0}{3}, \dfrac{-8}{2}, -10\right\}$ **2d.** $\left\{14, \dfrac{0}{3}, \dfrac{-8}{2}, -10, 2.\overline{6}\right\}$ **2e.** $\{-1.050050005...\}$

2f. $\left\{14, \dfrac{0}{3}, \dfrac{-8}{2}, -10, 2.\overline{6}, -1.050050005...\right\}$ **3a.** $Ir, \mathbb{R}$ **3b.** $\mathbb{Z}, \mathbb{Q}, \mathbb{R}$ **3c.** $\mathbb{Q}, \mathbb{R}$ **3d.** $\mathbb{N}, \mathbb{W}, \mathbb{Z}, \mathbb{Q}, \mathbb{R}$

4. See Graphing Answer Section **5a.** > **5b.** = **5c.** > **6a.** True **6b.** True **6c.** False **7a.** $x > 0$ **7b.** $x < 0$ **8.** positive **9a.** 0 **9b.** 4.2 **9c.** 120

Do the Math **1.** $B = \{1, 2, 3, 4, ..., 24\}$ **2.** $C = \{-5, -4, -3, -2, -1, 0, 1, 2, 3\}$ **3.** $F = \{ \ \}$ or $\varnothing$ **4.** 3, 0

5. $-4, 3, -\dfrac{13}{2}, 0$ **6.** All numbers listed **7.** See Graphing Answer Section **8.** False **9.** True **10.** > **11.** < **12.** =

13. 8 **14.** 7 **15.** $\dfrac{13}{9}$

Section 1.4

Five-Minute Warm-Up **1.** division **2.** subtraction **3.** addition **4.** multiplication **5a.** $\dfrac{9}{18}$ **5b.** 9/18

5c. $9 \div 18$ **5d.** $18\overline{)9}$ **6a.** 13 **6b.** $\dfrac{15}{4}$ **6c.** 0 **7.** $\dfrac{5}{2}$

Guided Practice **1a.** positive **1b.** negative **2.** the same as the sign of the integer having the larger absolute value. **3a.** 45 **3b.** 22 **3c.** 67 **3d.** negative **3e.** −67 **4a.** 73 **4b.** 81 **4c.** 8 **4d.** −81 **4e.** negative **4f.** negative

4g. −8 **5a.** $-\dfrac{9}{2}$ **5b.** 10.3 **6a.** $-32 + 61$ **6b.** 29 **7a.** positive **7b.** negative **7c.** zero **8a.** −72 **8b.** 60

8c. −117 **9a.** 36 **9b.** 12 **9c.** 3 **10a.** −72 **10b.** 72 **10c.** −1 **11.** $-\dfrac{1}{3}$

Do the Math 1. 8 2. –18 3. –213 4. 2 5. 34 6. –7 7. –7 8. –24 9. 71 10. 63 11. 110 12. –896 13. –192

14. $\dfrac{1}{10}$ 15. $-\dfrac{1}{3}$ 16. 4 17. –24 18. $\dfrac{20}{3}$ 19. $32 + (-64) = -32$ 20. $-40 \div 100$ or $\dfrac{-40}{100} = -\dfrac{2}{5}$

Section 1.5

Five-Minute Warm-Up 1. 90 2. –25 3. –78 4. –136 5. $\dfrac{1}{4}$ 6. 120 7. $\dfrac{25}{60}$

Guided Practice 1a. $-\dfrac{4}{5}$ 1b. –8 1c. $\dfrac{1}{5}$ 2. $-\dfrac{9}{40}$ 3a. $\dfrac{8}{15} \cdot \dfrac{45}{4}$ 3b. $\dfrac{2 \cdot 2 \cdot 2 \cdot 3 \cdot 3 \cdot 5}{2 \cdot 2 \cdot 3 \cdot 5}$

3c. $\dfrac{2 \cdot \cancel{2} \cdot \cancel{2} \cdot \cancel{3} \cdot 3 \cdot \cancel{5}}{\cancel{2} \cdot \cancel{2} \cdot \cancel{3} \cdot \cancel{5}}$ 3d. $\dfrac{6}{1} = 6$ 4a. $\dfrac{7-11}{12}$ 4b. $\dfrac{7+(-11)}{12}$ 4c. $\dfrac{-4}{12}$ 4d. $-\dfrac{1}{3}$ 5a. $2 \cdot 2 \cdot 3$

5b. $3 \cdot 3$ 5c. $2 \cdot 2 \cdot 3 \cdot 3$ 5d. 36 5e. $\dfrac{3}{3}$ 5f. $\dfrac{4}{4}$ 5g. $\dfrac{5}{12} = \dfrac{15}{36}; \dfrac{4}{9} = \dfrac{16}{36}$ 5h. $\dfrac{15+16}{36}$ 5i. $\dfrac{31}{36}$ 6a. $\dfrac{-3}{1}$

6b. $\dfrac{-24}{8} + \dfrac{5}{8}$ 6c. $\dfrac{-19}{8}$ 6d. $-\dfrac{19}{8}$ 7. at the right end of the integer. 8. sum

Do the Math 1. $\dfrac{5}{12}$ 2. –27 3. $-\dfrac{8}{5}$ 4. $\dfrac{5}{9}$ 5. $\dfrac{4}{9}$ 6. $-\dfrac{1}{8}$ 7. 1 8. $-\dfrac{1}{16}$ 9. $-\dfrac{40}{27}$ 10. $\dfrac{4}{3}$ 11. 2 12. $\dfrac{25}{8}$

13. $-\dfrac{16}{15}$ 14. $-\dfrac{203}{60}$ 15. 43.2 16. 26.32 17. 33.79 18. 22.1 19. 42.1 20. 1600

Section 1.6

Five-Minute Warm-Up 1. –2 2. –12 3. $\dfrac{2}{3}$ 4. 45 5. 5

Guided Practice 1a. $\dfrac{4}{7}$ 1b. $-\dfrac{1}{4}$ 2a. $16\dfrac{1}{2}$ feet 2b. 900 seconds 3. order 4. division; subtraction

5a. $6 + (-12)$ 5b. $\dfrac{1}{2} \cdot (4 + 20)$ 6a. 4 6b. 90 7. grouping 8a. $-9 + (12 + 2)$ 8b. $\left(3 \cdot \dfrac{2}{3}\right) \cdot 15$ 9a. 20

9b. 50 10a. 0 10b. 0 10c. undefined

Do the Math 1. 43 and $\dfrac{1}{3}$ yards = 43 yards, 1 foot 2. 59 meters 3. 14 gallons, 2 quarts

4. 7 pounds, 8 ounces 5. Commutative Property of Multiplication 6. Associative Property of Multiplication

7. $\dfrac{a}{0}$ is undefined 8. Multiplicative Inverse Property 9. 59 10. 28 11. –72 12. 13 13. 0 14. 2 15. 1 16. 30

17. $-\dfrac{21}{16}$

Section 1.7

Five-Minute Warm-Up 1. –81 2. –48 3. –32 4. –45 5. 70 6. 81

Guided Practice 1a. base 1b. exponent 1c. $2 \cdot 2 \cdot 2 \cdot 2 \cdot 2 = 32$ 2a. 9 2b. –9 3a. squared

3b. $(-6)^2$ 3c. $(-6)(-6) = 36$ 4a. cubed 4b. $(-5)^3$ 4c. $(-5)(-5)(-5) = -125$ 5a. Parentheses

5b. Exponents 5c. Multiply or Divide 5d. Add or Subtract 6. [ ], { }, | | 7. –31 8a. $\dfrac{4 \cdot 2^3 + (-16)}{3(-2)^2}$

8b. $\dfrac{4 \cdot 8 + (-16)}{3(4)}$ 8c. $\dfrac{32 + (-16)}{12}$ 8d. $\dfrac{16}{12}$ 8e. $\dfrac{4}{3}$

Do the Math 1. 4^5 2. $(-8)^3$ 3. 32 4. $\dfrac{625}{16}$ 5. 0.0016 6. –625 7. 1 8. $-\dfrac{243}{32}$ 9. 36 10. 40 11. 5 12. $\dfrac{4}{9}$

13. 23 14. 13.5 15. –26 16. $-\dfrac{49}{2}$ 17. $\dfrac{81}{64}$ 18. 24 19. $\dfrac{5}{24}$

Five-Minute Warm-Up **1.** 23 **2.** −10 **3.** −45 **4.** 7 **5a.** 16 **5b.** −16 **6.** 225

Guided Practice **1.** variable **2.** constant **3.** algebraic expression **4.** evaluating **5.** 21 **6.** term

7. coefficient **8.** like terms **9.** x^2; $-2xy$; $-y^2$ **10a.** $\dfrac{1}{2}$ **10b.** −1 **10c.** 14 **10d.** $-\dfrac{2}{5}$ **11a.** unlike **11b.** like

12a. $6x - 2$ **12b.** $-2y - 1$ **13a.** $4x$ **13b.** $6x + 3$ **14.** $9x + 12$

Do the Math **1.** −1 **2.** −32 **3.** $3m^4$, $-m^3n^2$, $4n$, −1; 3, −1, 4, −1 **4.** t^3, $-\dfrac{t}{4}$; 1, $-\dfrac{1}{4}$ **5.** like **6.** unlike

7. $12s + 6$ **8.** $-5k + 5n$ **9.** $6x + 9y$ **10.** $-5p^5$ **11.** $6m - 9n + 8p$ **12.** $9m - 6$ **13.** $\dfrac{13}{10}y$ **14.** $-21x + 21$ **15.** $884

Chapter 2 Answers
Section 2.1

Five-Minute Warm-Up **1.** $\dfrac{2}{3}$ **2.** $\dfrac{1}{14}$ **3.** 1 **4.** $-18x - 15$ **5.** $11 - 6x$ **6.** −26

Guided Practice **1.** Substitute the value into the original equation. If this results in a true statement, then the value for the variable is a solution to the equation. **2a.** Yes **2b.** No **3.** Whatever you add to one side of the equation, you must add to the other side of the equation to form an equivalent equation. **4.** Yes. Since $a - b = a + (-b)$, subtraction is an alternate way of writing addition. Subtraction means to add the opposite.

5. isolating the variable **6a.** $x + 13 - 13 = -11 - 13$ **6b.** $x + 0 = -24$ **6c.** $x = -24$ **6d.** $\{-24\}$

7. When you multiply one side of an equation by a non-zero quantity, you must also multiply the other side by the same non-zero quantity to form an equivalent equation. Yes, this property applies to division. Division means to multiply by the reciprocal. **8a.** $-\dfrac{1}{9} \bullet (-9x) = -\dfrac{1}{9} \bullet (324)$ **8b.** $\left(-\dfrac{1}{9} \bullet (-9)\right)x = -\dfrac{1}{9} \bullet 324$

8c. $1 \bullet x = -36$ **8d.** $x = -36$ **8e.** $\{-36\}$ **9a.** Divide both sides by 3; which is the same as multiply both sides by $\dfrac{1}{3}$. **9b.** Multiply both sides by 15 **9c.** Multiply both sides by $-\dfrac{3}{2}$ **9d.** Multiply both sides by $\dfrac{9}{4}$

Do the Math **1.** no **2.** no **3.** yes **4.** no **5.** 19 **6.** −15 **7.** $\dfrac{1}{2}$ **8.** $\dfrac{13}{24}$ **9.** $\dfrac{15}{2}$ **10.** $\dfrac{-5}{2}$ **11.** 12 **12.** 30 **13.** 14

14. $-\dfrac{5}{9}$ **15.** $799

Section 2.2

Five-Minute Warm-Up **1.** $-11x - 8$ **2.** 93 **3.** x **4.** $2x - 9$ **5.** $\dfrac{3}{7}$ **6.** 132.5

Guided Practice **1a.** 9 **1b.** $6x - 9 + 9 = -27 + 9$ **1c.** $6x = -18$ **1d.** 6; $\dfrac{1}{6}$ **1e.** $\dfrac{6x}{6} = \dfrac{-18}{6}$ **1f.** $x = -3$

1g. $\{-3\}$ **2a.** $\dfrac{5}{4}n + 3 - 3 = -12 - 3$ **2b.** $\dfrac{5}{4}n = -15$ **2c.** $\dfrac{4}{5}\left(\dfrac{5}{4}n\right) = \dfrac{4}{5}(-15)$ **2d.** $n = -12$ **2e.** $\{-12\}$

3. combining like terms on each side of the equation. **4.** Distributive **5a.** $-2x + 5 = -4$

5b. $24n + 36 = -2n - 1$ **6.** Add $5x$ to each side of the equation, and then subtract 4 from each side. Answers may vary. **7a.** $3z + 15 - 16z = 2 - z - 3$ **7b.** $-13z + 15 = -z - 1$

7c. $-13z + z + 15 = -z + z - 1$ **7d.** $-12z + 15 = -1$ **7e.** $-12z + 15 - 15 = -1 - 15$ **7f.** $-12z = -16$

7g. $\dfrac{-12z}{-12} = \dfrac{-16}{-12}$ **7h.** $z = \dfrac{4}{3}$ **7i.** $\left\{\dfrac{4}{3}\right\}$

Do the Math **1.** −3 **2.** $\dfrac{5}{3}$ **3.** 8 **4.** 25 **5.** −8 **6.** −3 **7.** −2 **8.** −2 **9.** 2 **10.** $\dfrac{1}{3}$ **11.** 2 **12.** $\dfrac{8}{3}$ **13.** 35 g **14.** $8.75

15. 68 inches, 70 inches, 72 inches

Section 2.3

Five-Minute Warm-Up **1.** 35 **2.** 225 **3.** $-3x + 4$ **4.** $43x - 270$ **5.** $x + 28$ **6.** $4x - 8$ **7.** $10x - 10$

Guided Practice **1a.** 12 **1b.** $12 \cdot \left(\frac{3}{2}x - \frac{4}{3}x\right) = 12 \cdot \left(-\frac{9}{4}\right)$ **1c.** $12 \cdot \left(\frac{3}{2}x\right) - 12 \cdot \left(\frac{4}{3}x\right) = 12 \cdot \left(-\frac{9}{4}\right)$

1d. $18x - 16x = -27$ **1e.** $2x = -27$ **1f.** $\frac{2x}{2} = \frac{-27}{2}$ **1g.** $x = -\frac{27}{2}$ **1h.** $\left\{-\frac{27}{2}\right\}$ **2a.** 18

2b. $18\left(\frac{4x+3}{9} + 1\right) = 18\left(\frac{2x+1}{2}\right)$ **2c.** $2(4x+3) + 18 \cdot 1 = 9(2x+1)$ **2d.** $8x + 6 + 18 = 18x + 9$

2e. $8x + 24 = 18x + 9$ **3.** $\{3000\}$ **4a.** false **4b.** $\varnothing$ **4c.** contradiction **5a.** 81 **5b.** yes; yes **5c.** 46; 81

Do the Math **1.** 3 **2.** $\frac{44}{9}$ **3.** 8 **4.** $\frac{8}{3}$ **5.** 20 **6.** 250 **7.** 8 **8.** 1 **9.** $\varnothing$ or $\{\ \}$; contradiction

10. all real numbers; identity **11.** $\left\{\frac{3}{2}\right\}$; conditional equation **12.** $\varnothing$ or $\{\ \}$; contradiction **13.** $48

14. 10 nickels **15.** $86,200

Section 2.4

Five-Minute Warm-Up **1.** 28 **2.** 12.38 **3.** 0.0725 **4.** 78.5 **5.** $\{4\}$

Guided Practice **1.** $68°F$ **2.** *I:* interest earned; *r:* annual interest rate, as a decimal;
P: Principal (amount borrowed or amount deposited); *t* time of deposit, typically in years **3a.** $A = s^2$; $P = 4s$

3b. $A = lw$; $P = 2l + 2w$ **3c.** $A = \frac{1}{2}bh$; $P = a + b + c$ **3d.** $A = \frac{1}{2}h(B + b)$; $P = a + b + c + B$

3e. $A = ah$; $P = 2a + 2b$ **3f.** $A = \pi r^2$; $C = 2\pi r = \pi d$ **3g.** $V = s^3$; $S = 6s^2$

3h. $V = lwh$; $S = 2lw + 2lh + 2wh$ **3i.** $V = \frac{4}{3}\pi r^3$; $S = 4\pi r^2$ **3j.** $V = \pi r^2 h$; $S = 2\pi r^2 + 2\pi rh$

3k. $V = \frac{1}{3}\pi r^2 h$ **4a.** $P = 2l + 2w$ **4b.** 30.5 m **4c.** $457.50 **5a.** $6x - 6x - 12y = -6x - 24$

5b. $-12y = -6x - 24$ **5c.** $\frac{-12y}{-12} = \frac{-6x - 24}{-12}$ **5d.** $y = \frac{1}{2}x + 2$ **6a.** $C = R - P$ **6b.** $C = 2100

Do the Math **1.** 1066.8 meters **2.** $834 **3.** $150 **4a.** 104 units **4b.** 640 units^2 **5a.** 17.58 units

5b. 24.62 units^2 **6.** $m = \frac{F}{v^2}$ **7.** $B = \frac{3V}{h}$ **8.** $b = S - a - c$ **9.** $l = \frac{P - 2w}{2}$ **10.** $y = 2x + 18$ **11.** $y = 2x - 3$

12. $y = -\frac{5}{6}x + 3$ **13.** $y = \frac{4}{15}x - 2$ **14a.** $R = P + C$ **14b.** $6000 **15a.** $t = \frac{I}{Pr}$ **15b.** 2 years

Section 2.5

Five-Minute Warm-Up **1.** $\{-54.72\}$ **2.** $\{32\}$ **3a.** $\times$ **3b.** $-$ **3c.** $\times$ **3d.** $+$ **3e.** $\div$ **3f.** $+$ **3g.** $-$
3h. $\div$ **3i.** $-$ **3j.** $+$ **3k.** $\div$ **3l.** $-$ **3m.** $\times$ **3n.** $+$ **3o.** $+$ **3p.** $\times$ **3q.** $-$ **3r.** $\times$ **3s.** $\times$ **3t.** $+$

Guided Practice **1a.** $-5 + 3x$ or $3x - 5$ **1b.** $5(r + 25)$ **1c.** $3x + 12$ **1d.** $\frac{2p}{7}$ **1e.** $\frac{z}{2} - 18$ or $\frac{1}{2}z - 18$

1f. $w - 900$ **2a.** $b - a$ **2b.** $a - b$ **2c.** $b - a$ **2d.** $\frac{a}{b}$ **2e.** $\frac{b}{a}$ **3a.** $x + 5 = 19$ **3b.** $2y - 14 = 30$

3c. $z - 9 = 6 + \frac{2}{5}z$ **4.** Identify what you are looking for; Give names to the unknowns; Translate the problem

into the language of mathematics; Solve the equation(s) found in Step 3; Check the reasonableness of your
answer; Answer the question **5a.** $n + 1$ **5b.** $n + 2$ **6a.** $x + 2$ **6b.** $x + 4$ **6c.** $x + 6$ **7a.** $p + 2$ **7b.** $p + 4$
7c. $p + 6$ **8a.** $n + 2$; $n + 4$ **8b.** $n + (n + 2) + (n + 4) = 453$ **8c.** $n = 149$ **8d.** 149, 151, 153 are the three

odd integers **9a.** $x - 5000$ **9b.** $x + (x - 5000) = 20,000$ **9c.** $x = 12,500$ **9d.** $12,500 in stocks and $7500 in
bonds

Do the Math **1.** $2x$ **2.** $x - 8$ **3.** $\dfrac{-14}{x}$ **4.** $4x + 21$ **5.** Indians: r; Blue Jays: $r - 3$ **6.** Ralph's amount: R; Beryl's amount: $3R + 0.25$ **7.** number paid tickets: p; number promotion tickets: $12{,}765 - p$ **8.** $43 + x = -72$

9. $49 = 2x - 3$ **10.** $\dfrac{x}{-6} - 15 = 30$ **11.** $2(x + 5) = x + 7$ **12.** 25, 27, 29 **13.** Ryugyong Hotel: 105 stories;

Sears Tower: 110 stories **14.** Stocks: \$24,000; bonds: \$16,000 **15.** 240 minutes

Section 2.6

Five-Minute Warm-Up **1a.** 0.62 **1b.** 0.0175 **2a.** 5.5% **2b.** 150% **3.** $1.75p$ **4.** $\{13.25\}$

Guided Practice **1a.** $\dfrac{3}{4}$ **1b.** 1 **1c.** $\dfrac{21}{400}$ **2a.** 0.68 **2b.** 0.025 **2c.** 1.5 **3a.** 8% **3b.** 70% **3c.** 30%

4a. $n = 0.72(30)$ **4b.** 21.6 **5a.** $150 = x \bullet (120)$ **5b.** 125% **6a.** $0.075p$

6b. $p + 0.075p = 1343.75$ or $1.075p = 1343.75$ **6c.** $p = 1250$ **6d.** The computer was \$1250 before sales tax.

7a. $0.60p$ **7b.** $p - 0.6p = 224$ or $0.4p = 224$ **7c.** $p = 560$ **7d.** The price of the skis, before the discount, was \$560.

Do the Math **1.** 42.5 **2.** 15 **3.** 13.5 **4.** 305 **5.** 2000 **6.** 200 **7.** 16% **8.** 75% **9.** \$8000 **10.** \$22.57 **11.** \$200,000 **12.** \$15,550 **13.** \$41.48 **14.** \$1627 **15.** 24.86 million

Section 2.7

Five-Minute Warm-Up **1.** $\{50\}$ **2.** $\left\{\dfrac{7}{2}\right\}$ **3a.** $12 + 2x$ **3b.** $\dfrac{x}{2} - 25$ **3c.** $2(45 - x)$ **3d.** $15 + 2x$

3e. $2(x + 30)$ **4.** $8 - p$

Guided Practice **1.** $90°$ **2.** $180°$ **3a.** $x - 20$ **3b.** $x + x - 20 = 180$ **3c.** $x = 100$ **3d.** The large angle is $100°$ and the smaller angle is $80°$. **4.** $180°$ **5a.** $2x$ **5b.** $30 + 2x$ **5c.** $x + 2x + 2x + 30 = 180$ **5d.** $x = 30$ **5e.** The first angle measures $30°$, the second angle measures $60°$, and the third angles measures $90°$.

6a. $x + 20$ **6b.** $P = 2l + 2w$ **6c.** $200 = 2(x + 20) + 2x$ **6d.** $x = 40$; The dimensions of the tablecloth are 40" by 60". **7a.** The distances add to 50 miles. **7b.** See Graphing Answer Section

7c. $2.5r + 2.5(r + 4) = 50$; $r = 8$ **7d.** The slower boat is traveling at 8 mph and the faster boat is traveling at 12 mph.

Do the Math **1.** $x = 57.5$; $57.5°$ and $32.5°$ **2.** $18°$, $72°$, $90°$ **3.** length = 24 meters; width = 2 meters **4.** 7.5 hr **4a.** $72t$ **4b.** $66t$ **4c.** $72t - 66t$ **4d.** $72t - 66t = 45$ **5.** See Graphing Answer Section; $2r + 3(r - 10) = 580$ **6.** 4 feet **7.** 12 feet **8.** length = 54 inches; width = 36 inches **9.** eastbound: 18 mph

10. $2\dfrac{1}{2}$ hours

Section 2.8

Five-Minute Warm-Up **1.** $<$ **2.** $<$ **3.** $>$ **4.** $=$ **5.** $>$ **6.** $<$ **7.** See Graphing Answer Section

Guided Practice **1.** square bracket; parenthesis **2.** See Graphing Answer Section **3a.** $(-\infty, -5)$

3b. $[10, \infty)$ **4.** true **5.** true **6.** false **7a.** $3x + 10 - 10 > 1 - 10$ **7b.** $3x > -9$ **7c.** $x > -3$ **7d.** $\{x \mid x > -3\}$

7e. $(-3, \infty)$ **7f.** See Graphing Answer Section **8a.** $(12, \infty)$ **8b.** $(-\infty, -25]$

9a. $6 \bullet \dfrac{1}{6}(4x - 5) \le 6 \bullet \dfrac{1}{2}(2x + 3)$ **9b.** $1 \bullet (4x - 5) \le 3 \bullet (2x + 3)$ **9c.** $4x - 5 \le 6x + 9$ **9d.** $-2x - 5 \le 9$

9e. $-2x \le 14$ **9f.** $x \ge -7$ **9g.** $\{x \mid x \ge -7\}$ **9h.** $[-7, \infty)$ **10a.** $\ge$ **10b.** $\ge$ **10c.** $>$ **10d.** $>$ **10e.** $\le$ **10f.** $\le$

10g. $<$ **10h.** $<$

Do the Math **1.** See Graphing Answer Section; $(5, \infty)$ **2.** See Graphing Answer Section; $(-\infty, 6]$

3. See Graphing Answer Section; $[-2, \infty)$ **4.** See Graphing Answer Section; $(-\infty, -3)$ **5.** $(-\infty, 3]$ **6.** $(-2, \infty)$

7. $>$; Addition Property of Inequality **8.** $\ge$; Multiplication Property of Inequality **9.** $\{x \mid x < 3\}$; $(-\infty, 3)$; See Graphing Answer Section **10.** $\{x \mid x > 3\}$; $(3, \infty)$; See Graphing Answer Section **11.** $\{x \mid x \le -4\}$; $(-\infty, -4]$; See Graphing Answer Section **12.** $\{x \mid x > -2\}$; $(-2, \infty)$; See Graphing Answer Section **13.** $\{x \mid x \ge 5\}$; $[5, \infty)$; See Graphing Answer Section **14.** $\{x \mid x \ge -1\}$; $[-1, \infty)$; See Graphing Answer Section

15. $\left\{x \mid x > \dfrac{11}{2}\right\}; \left(\dfrac{11}{2}, \infty\right)$; See Graphing Answer Section **16.** $\varnothing$ or $\{\ \}$; See Graphing Answer Section

17. $\{p \mid p \text{ is any real number}\}; (-\infty, \infty)$; See Graphing Answer Section **18.** $x \geq 250$ **19.** $x < 25$
20. at most 71 miles

Chapter 3 Answers
Section 3.1

Five-Minute Warm-Up **1.** See Graphing Answer Section **2a.** 7 **2b.** -3 **3a.** -25 **3b.** 9 **4.** $\{5\}$ **5.** $\{3\}$

Guided Practice **1.** See Graphing Answer Section **2.** the origin **3.** left **4A** $(4, -3)$ **4B** $(-3, -1)$

4C $(-2, 1)$ **5a.** No **5b.** Yes **6a.** $7 - 12 = -5$ **6b.** $-\dfrac{5}{2}$ **6c.** $\left(3, -\dfrac{5}{2}\right)$ **7a.** $-2; -7; (-2, -7)$

7b. $-1; -4; (-1, -4)$ **7c.** $0; -1; (0, -1)$ **7d.** $1; 2; (1, 2)$ **7e.** $2; 5; (2, 5)$

Do the Math **1.** See Graphing Answer Section; Quadrant I: R; Quadrant II: S; Quadrant III: P, T; Quadrant IV: Q, U **2.** See Graphing Answer Section; Quadrant I: U; Quadrant III: T; Quadrant IV: P, V; x-axis: R, S; y-axis: Q, S **3.** $P(0, 2)$: y–axis; $Q(-3, -1)$: Quadrant III; $R(-2, 4)$: Quadrant II; $S(2, 2)$: Quadrant I; $T(3, -2)$: Quadrant IV; $U(-5, 0)$: x–axis **4.** A Yes B No C Yes **5.** A No B No C Yes **6.** $(2, 5)$ **7.** $(4, 3)$

8.–9. See Graphing Answer Section **10.** A -9 B 7 C -6 **11.** A $\dfrac{1}{6}$ B $\dfrac{3}{2}$ C $-\dfrac{1}{2}$ **12a.** \$11.70 **12b.** \$41.70 **12c.**

15.5 mi **12d.** It costs \$29.70 to take a taxi 14 miles.

Section 3.2

Five-Minute Warm-Up **1.** $\{-9\}$ **2.** $\left\{-\dfrac{10}{3}\right\}$ **3.** $\{-2\}$ **4a.** 22 **4b.** 12 **5a.** $y = -1$ **5b.** $x = -4$

6. See Graphing Answer Section
Guided Practice **1.** See Graphing Answer Section **2.** standard **3.** $y = 0$ **4.** $x = 0$ **5a.** -2 **5b.** 3
5c.–7. See Graphing Answer Section
Do the Math **1.** Not linear **2.** Linear **3.–4.** See Graphing Answer Section **5.** $(0, 2), (2, 0)$ **6.** $(0, -3), (6, 0)$ **7.** $(0, 6)$ **8.** $(7, 0)$ **9.** $(0, -6), (10, 0)$ **10.** $(0, -8), (2, 0)$ **11.–14.** See Graphing Answer Section

Section 3.3

Five-Minute Warm-Up **1.** $\dfrac{1}{2}$ **2.** 3 **3.** $-\dfrac{5}{3}$ **4.** -2 **5.** undefined **6.** 0 **7.** $-\dfrac{5}{6}$

Guided Practice **1.** slope **2.** $m = \dfrac{y_2 - y_1}{x_2 - x_1}$ **3.** -5 **4.** undefined **5.** vertical **6.** 0 **7.** horizontal **8a.** slants

upward from left to right **8b.** slants downward from left to right **8c.** horizontal **8d.** vertical
9. See Graphing Answer Section

Do the Math **1.** $\dfrac{3}{4}$ **2.** $-\dfrac{7}{4}$ **3a., 3b.** See Graphing Answer Section **3c.** $m = \dfrac{5}{2}$; the value of y increases by 5

units when x increases by 2 units. **4a., 4b.** See Graphing Answer Section **4c.** $m = -\dfrac{1}{6}$; the value of y decreases

by 1 unit when x increases by 6 units. **5.** $m = 0$ **6.** m is undefined **7.–8.** See Graphing Answer Section **9.** $\dfrac{4}{45}$

10. $m = 0.225$ million; the income increases by 0.225 million dollars or \$225,000 for each year of college

Section 3.4

Five-Minute Warm-Up **1.** $y = -\dfrac{7}{5}x - 7$ **2.** $y = \dfrac{3}{8}x - \dfrac{3}{2}$ **3.** $\left\{-\dfrac{4}{3}\right\}$ **4a.** $-\dfrac{5}{3}$ **4b.** $-\dfrac{1}{2}$

Guided Practice **1.** $m; b$ **2.** $-\dfrac{1}{2}; 6$ **3.** $-2; 5$ **4a.** $\dfrac{2}{5}$ **4b.** -1 **4c.** See Graphing Answer Section

5a. $y = -\dfrac{2}{3}x + 2$ **5b.** $-\dfrac{2}{3}$ **5c.** 2 **5d.** See Graphing Answer Section **6.** $y = -\dfrac{1}{4}x + 3$ **7a.** $y = 0.07x + 22$

7b. \$29

Do the Math 1. $m = -6; b = 2$ **2.** $m = -1; b = -12$ **3.** $m = \dfrac{3}{4}; b = 3$ **4.** $m = -\dfrac{5}{3}; b = 4$ **5.– 8.** See Graphing

Answer Section **9.** $y = x + 10$ **10.** $y = \dfrac{4}{7}x - 9$ **11.** $y = \dfrac{1}{4}x + \dfrac{3}{8}$ **12.** $y = -2$ **13.** $x = 4$ **14.** $y = -3x$

15a. 2625 calories **15b.** 9 years old **15c.** The recommended caloric intake increases 125 calories per year.
15d. Age 3 is outside the given range this equation applies to. **15e.** See Graphing Answer Section

Section 3.5

Five-Minute Warm-Up 1. $y = -4x + 10$ **2.** $y = 3x + 4$ **3.** $-\dfrac{1}{3}$ **4.** -2 **5.** $-7x + 21$ **6.** -1

Guided Practice 1. $y - y_1 = m(x - x_1)$ **2a.** $9; -2; -1$ **2b.** $y - y_1 = m(x - x_1)$

2c. $y - (-1) = 9(x - (-2))$ **2d.** $y + 1 = 9(x + 2)$ **2e.** $y + 1 = 9x + 18$ **2f.** $y = 9x + 17$ **3a.** $-\dfrac{5}{6}; 0; 12$

3b. $y - y_1 = m(x - x_1)$ **3c.** $y - (12) = -\dfrac{5}{6}(x - 0)$ **3d.** $y = -\dfrac{5}{6}x + 12$ **4a.** $0; 5; -11$

4b. $y - y_1 = m(x - x_1)$ **4c.** $y - (-11) = 0(x - 5)$ **4d.** $y + 11 = 0; y = -11$ **5.** the slope **6a.** $m = \dfrac{y_2 - y_1}{x_2 - x_1}$

6b. $x_1 = 1, y_1 = 5, x_2 = -1, y_2 = -1$ **14c.** $m = \dfrac{-1 - 5}{-1 - 1} = \dfrac{-6}{-2} = 3$ **6d.** $y - y_1 = m(x - x_1)$

6e. $m = 3, x_1 = 1, y_1 = 5$ **6f.** $y - 5 = 3(x - 1)$ **6g.** $y - 5 = 3x - 3$ **6h.** $5; y = 3x + 2$ **6i.** yes **6j.** no

6k. 3 **6l.** 2 **7a.** $\dfrac{3 + 1}{-5 + 5} = \dfrac{4}{0} =$ undefined **7b.** vertical **7c.** -5 **7d.** $x = -5$

Do the Math 1. $y = 6x - 23$; See Graphing Answer Section **2.** $y = -5x + 27$; See Graphing Answer Section

3. $y = -\dfrac{1}{2}x - 2$; See Graphing Answer Section **4.** $y = -1$; See Graphing Answer Section **5.** $y = -1$ **6.** $x = 4$

7. $y = -\dfrac{3}{2}x + 1$ **8.** $y = 3x + 6$ **9.** $y = 5$ **10.** $x = -3$ **11a.** After 30 years, monthly income is $3600.

11b. See Graphing Answer Section **11c.** $y = 45x + 2250$ **11d.** $2925 **11e.** Monthly income increases by $45 for each additional year of service.

Section 3.6

Five-Minute Warm-Up 1. $-\dfrac{1}{5}$ **2.** 3 **3.** $-\dfrac{4}{9}$ **4.** $y = 4x - 12$ **5a.** $\dfrac{1}{4}$ **5b.** $-\dfrac{1}{2}$ **6.** $y = -\dfrac{3}{2}x + \dfrac{11}{2}$

Guided Practice 1. Two lines are parallel if they never intersect. **2.** slope; y-intercept **3.** $=; \neq$
4a. $y = 4x - 1$ **4b.** $4; -1$ **4c.** $y = 4x + 5$ **4d.** $4; 5$ **4e.** Since the slopes are equal and the y-intercepts are

different, we know that L_1 is parallel to L_2 **5a.** $y = \dfrac{5}{2}x + \dfrac{1}{2}$ **5b.** $\dfrac{5}{2}$ **5c.** $\dfrac{5}{2}$ **5d.** $\dfrac{5}{2}; -3; -4$

5e. $y - y_1 = m(x - x_1)$ **5f.** $y - (-4) = \dfrac{5}{2}(x - (-3))$ **5g.** $y + 4 = \dfrac{5}{2}x + \dfrac{15}{2}; y = \dfrac{5}{2}x + \dfrac{7}{2}$

6. Two lines are perpendicular if they intersect at right angles $(90°)$. **7.** -1 **8.** $-1; m_1 = -\dfrac{1}{m_2}$ **9a.** $\dfrac{3}{4}; 6$

9b. $\dfrac{4}{3}; -2$ **9c.** $\dfrac{3}{4} \cdot \dfrac{4}{3} = 1$ **9d.** Since the product of the slopes is not -1, L_1 is not perpendicular to L_2.

10a. $y = 2x + 7$ **10b.** 2 **10c.** $-\dfrac{1}{2}$ **10d.** $-\dfrac{1}{2}; 3; -12$ **10e.** $y - y_1 = m(x - x_1)$

10f. $y - (-12) = -\dfrac{1}{2}(x - (3))$ **10g.** $y + 12 = -\dfrac{1}{2}x + \dfrac{3}{2}; y = -\dfrac{1}{2}x - \dfrac{21}{2}$

Do the Math **1-3.** See Graphing Answer Section **4.** neither **5.** perpendicular **6.** parallel **7.** neither

8. $y = 2x - 19$ **9.** $x = -4$ **10.** $y = -8$ **11.** $y = -\frac{2}{3}x + 11$ **12.** $y = -3x + 19$ **13.** $y = \frac{1}{2}x - 4$ **14.** $y = -6$

15. $x = 11$

Section 3.7

Five-Minute Warm-Up **1.** $\{x \mid x < -8\}$ **2.** $\{x \mid x \geq -3\}$ **3.** $\{x \mid x \leq -5\}$ **4.** See Graphing Answer Section

Guided Practice **1.** True **2.** yes **3.** dashed; solid **4.** half-planes **5.** shade **6a.** $y = x + 2$ **6b.** 1 **6c.** 2

6d. solid **6e.** $(0, 0)$ **6f.** false **6g.** opposite **6h.** See Graphing Answer Section **7a.** dashed **7b.** no

7c. See Graphing Answer Section **8a.** $4.32x + 2.50y \geq 5000$ **8b.** no

Do the Math **1.** A No B Yes C No **2.** A No B No C Yes **3.** A Yes B Yes C Yes **4.** A No B Yes C No
5.–12. See Graphing Answer Section **13a.** $160a + 75c \leq 500$ **13b.** Yes **13c.** Yes

Chapter 4 Answers
Section 4.1

Five-Minute Warm-Up **1a.–b.** See Graphing Answer Section **2.** Yes **3.** No

Guided Practice **1a.** No **1b.** Yes **2a.** –5 **2b.** 5 **2c.** $y = -2x - 1$ **2d.** –2 **2e.** –1 **2f.** $(-2, 3)$

2g. See Graphing Answer Section **2h.** $(-2, 3)$ **3.** See Graphing Answer Section; $\emptyset$ **4a.** inconsistent; the
lines are parallel **4b.** consistent and dependent; the lines are coincident **4c.** consistent and independent; the
lines intersect **5a.** infinitely many **5b.** consistent **5c.** dependent

Do the Math **1a.** Yes **1b.** No **1c.** No **2a.** No **2b.** No **2c.** Yes **3.** See Graphing Answer Section **4.** See
Graphing Answer Section; no solution; { } or $\emptyset$ **5.** See Graphing Answer Section **6.** See Graphing Answer
Section; infinitely many solutions **7.** no solution; inconsistent **8.** infinitely many solutions; consistent;
dependent **9.** one solution; consistent; independent **10.** infinitely many solutions; consistent; dependent

11. See Graphing Answer Section; $\begin{cases} y = 60 + 0.14x \\ y = 30 + 0.2x \end{cases}$ The cost for driving 500 miles is the same for both

companies ($130). Choose Slow-but-Cheap to drive 400 miles.

Section 4.2

Five-Minute Warm-Up **1.** $y = \frac{5}{3}x + 6$ **2.** $x = -\frac{8}{3}y - 4$ **3.** $\{-7\}$ **4.** $\{3\}$

Guided Practice **1a.** $4x + 3y = -23$ **1b.** $4x + 3(-x - 7) = -23$ **1c.** $4x - 3x - 21 = -23$

1d. $x - 21 = -23$ **1e.** $x = -2$ **1f.** $y = -x - 7$ **1g.** $y = -(-2) - 7$ **1h.** $y = -5$ **1i.** $(-2, -5)$ **2.** $(4, 10)$

3. infinitely many solutions; consistent/dependent **4a.** $x + y = -11$ **4b.** $x - y = -101$ **4c.** $\begin{cases} x + y = -11 \\ x - y = -101 \end{cases}$

4d. $(-56, 45)$ **4e.** The two numbers are -56 and 45.

Do the Math **1.** $(-3, 5)$ **2.** $(18, 13)$ **3.** $\left(-\frac{2}{5}, -5\right)$ **4.** $\left(-\frac{1}{2}, -2\right)$ **5.** no solution; $\emptyset$ or { }; inconsistent

6. infinitely many solutions; consistent and dependent **7.** no solution; $\emptyset$ or { }; inconsistent

8. no solution; $\emptyset$ or { }; inconsistent **9.** $\left(2, \frac{3}{2}\right)$ **10.** infinitely many solutions **11.** $\left(\frac{19}{21}, -\frac{6}{7}\right)$

12. no solution; $\emptyset$ or { } **13.** $\left(-\frac{1}{3}, \frac{1}{6}\right)$ **14.** infinitely many solutions

15. $8000 in international stock fund; $16,000 in domestic growth fund

Section 4.3

Five-Minute Warm-Up **1.** $-\frac{5}{4}$ **2.** 6 **3.** $21x - 14y$ **4.** $\emptyset$ **5.** $-9y - 9$ **6a.** 24 **6b.** 3 **6c.** 2

Sullivan/Struve/Mazzarella, *Elementary Algebra*, 2e

Guided Practice **1.** $(-5, 7)$ **2a.** $(1) -10x - 5y = 20$; (2) $3x + 5y = 29$ **2b.** $-7x = 49$ **2c.** $x = -7$
2d. $2(-7) + y = -4$; $y = 10$ **2e.** $(-7, 10)$ **3a.** $5x - 2y = 4$ **3b.** $-3x + 10y = -20$ **3c.** $(0, -2)$ **4.** $\varnothing$
5. infinitely many solutions **6a.** $a + c = 525$ **6b.** $7.50a + 4.50c = 3337.50$ **6c.** $(325, 200)$ **6d.** 325 adult
tickets and 200 children's tickets were sold.

Do the Math **1.** $(-3, 2)$ **2.** $\left(\dfrac{4}{3}, \dfrac{3}{2}\right)$ **3.** no solution; $\{\ \}$ or $\varnothing$; inconsistent

4. infinitely many solutions; dependent **5.** no solution; $\{\ \}$ or $\varnothing$ **6.** $(-14, 10)$ **7.** infinitely many solutions
8. $\left(\dfrac{95}{29}, -\dfrac{43}{29}\right)$ **9.** $\left(5, -\dfrac{1}{3}\right)$ **10.** $(-0.1, -0.2)$ **11.** $(1, 4)$ **12.** $(8, 17)$ **13.** biscuit 28 g; orange juice 42 g

Section 4.4

Five-Minute Warm-Up **1a.** 325 miles **1b.** 3 hours **2a.** \$945 **2b.** \$52.50

3a. $2x - 7 = 45$ **3b.** $5(x + 3) = \dfrac{x}{2}$ **3c.** $\dfrac{x}{3} = 20 - x$

Guided Practice **1a.** $4x + y = 68$ **1b.** $x - 2y = -1$ **1c.** $(15, 8)$ **1d.** The numbers are 15 and 8.
2. $180°$ **3a.** $x + y = 180$ **3b.** $x = 30 + y$ **3c.** $(105, 75)$ **3d.** The two supplementary angles are

$105°$ and $75°$. **4a.** $a + w$ **4b.** $a - w$ **4c–d.** See Graphing Answer Section **4e.** $\begin{cases} 6a + 6w = 2400 \\ 8a - 8w = 2400 \end{cases}$

4f. $(350, 50)$ **4g.** The speed the wind is moving at is 50 mph, and the air speed of the plane is 350 mph
Do the Math **1.** $20 + 3a = 2b + 50$ **2.** $l = 3w - 8$ **3.** $4(a - w) = 1200$ **4.** 19, 36 **5.** 18, 14
6. length 36.25 ft; width 26.25 ft **7.** $25°, 65°$ **8.** $96°, 84°$ **9.** airplane 485 mph; wind 30 mph
10. still water rate 5 mph; current speed 1 mph **11.** Gabriella's horse 2 mph; Monica's horse 6 mph

Section 4.5

Five-Minute Warm-Up **1a.** \$30 **1b.** \$2030 **2a.** 0.4 **2b.** 0.0015 **3.** $(7, -5)$

Guided Practice **1a–c.** See Graphing Answer Section **1d.** $\begin{cases} a + s = 40 \\ 7.5a + 4s = 202 \end{cases}$ **2a–c.** See Graphing

Answer Section **2d.** $\begin{cases} d + q = 42 \\ 0.1d + 0.25q = 6.75 \end{cases}$ **2e.** There are 25 dimes and 17 quarters. **3a–c.** See Graphing

Answer Section **3d.** $\begin{cases} a + p = 50 \\ 6.5a + 4p = 300 \end{cases}$ **4a–c.** See Graphing Answer Section **4d.** $\begin{cases} x + y = 90 \\ 0.25x + 0.4y = 27 \end{cases}$

Do the Math **1.** See Graphing Answer Section $\begin{cases} b + n = 69 \\ 10b + 15n = 895 \end{cases}$ **2.** See Graphing Answer Section

3. $\begin{cases} s + c = 1700 \\ 0.0275s + 0.02c = 37.75 \end{cases}$ See Graphing Answer Section $\begin{cases} p + t = 40 \\ 5p + 2t = 3(40) \end{cases}$ **4.** $\begin{cases} a + c = 22 \\ 15a + 7c = 274 \end{cases}$

5. $\begin{cases} A + B = 2650 \\ 0.05A + 0.065B = 155 \end{cases}$ **6.** $\begin{cases} 8.60B + 5.75W = 143.45 \\ B = 2W - 2 \end{cases}$ **7.** adult \$4.50; child \$3.25
8. \$1500 in 4.5% account; \$3500 in 9% account **9.** 14 liters

Section 4.6

Five-Minute Warm-Up **1.** $(-\infty, -2]$ **2.** $(-1, \infty)$ **3a.** See Graphing Answer Section
3b. See Graphing Answer Section
Guided Practice **1.** (c) **2.** graphing **3.** dashed **4.** solid **5.** the opposite half-plane
6. intersection **7.** See Graphing Answer Section **8a.** $5x + 8y \leq 40$ **8b.** $y \geq 2x$ **8c.** $x \geq 0$; $y \geq 0$
8d. See Graphing Answer Section **8e.** No
Do the Math **1a.** Yes **1b.** No **1c.** No **2a.** No **2b.** No **2c.** Yes **3.–7a.** See Graphing Answer Section
7b. Yes **7c.** No

Chapter 5 Answers

Section 5.1

Five-Minute Warm-Up **1.** -1 **2.** $-2x^2 + 4$ **3.** $-10x - 6y + 9$ **4.** $-15x + 10$ **5.** $-x^2 y + 3xy^2$ **6.** 7

Guided Practice **1a.** A single term which is the product of a constant and a variable with whole number exponents. **1b.** The constant multiplied with the variable. **1c.** The degree is the same as the exponent on the variable. **1d.** The degree is the sum of the exponents on the variables. **2a.** A polynomial is the sum of two or more monomials. **2b.** The degrees of the terms are in descending order. **2c.** The degree of the polynomial is the same as the highest degree of any of the terms in the polynomial. **3a.** monomial **3b.** binomial **3c.** trinomial **3d.** polynomial **4.** Like terms have the same variables, raised to the same powers. **5.** $7x^3 - x^2 - 5x + 5$ **6.** $-x^2 y - 14x^2 y^2 + xy^2$ **7.** $9z^3 - 7z^2 + z - 6$ **8.** $-n^2 + 2n$ **9a.** 5 **9b.** 41 **10.** -7

Do the Math **1.** Yes; coefficient -1; degree 7 **2.** No **3.** No **4.** Yes; $p^5 - 3p^4 + 7p + 8$; degree 5; polynomial **5.** $7x^2 - x - 3$ **6.** $-10w^2 + 6w - 2$ **7.** $\frac{29}{24}b^2 - \frac{7}{15}b$ **8.** $-2a^2 - 3ab - 8b^2$ **9.** $2x^2 + 3x - 7$ **10.** $-2m^4 + 2m^2 + 7$ **11.** $\frac{7}{12}x^2 - \frac{25}{24}x - 6$ **12.** $-6m^2 n + 4mn - 1$ **13.** 10; 9; 9 **14.** 10; 2.125; $\frac{65}{32}$ **15.** 4 **16.** $-\frac{3}{2}$

Section 5.2

Five-Minute Warm-Up **1a.** $\left(\frac{3}{4}\right)^3$ **1b.** $(-5)^4$ **2a.** -9 **2b.** $\frac{27}{8}$ **2c.** 16 **2d.** $-\frac{64}{27}$ **2e.** -32 **2f.** -36 **3.** -12 **4.** $20 - 10x$

Guided Practice **1.** sum **2a.** r^8 **2b.** $m^3 n^5$ **3.** multiply **4a.** p^{18} **4b.** $\left(-x^{12}\right) = x^{12}$ **5.** positive **6.** negative **7.** raised to that power **8a.** $125x^3 y^6$ **8b.** $m^8 n^{12}$ **9a.** $-12x^{11}$ **9b.** q^7 **10a.** $10m^5 n^7$ **10b.** $48p^4 q^{10}$

Do the Math **1.** 81 **2.** a^{10} **3.** b^{13} **4.** $(-z)^6$ **5.** 81 **6.** $(-n)^{21}$ **7.** k^{24} **8.** $(-a)^{18}$ **9.** $16y^6$ **10.** $\frac{27}{64}n^6$ **11.** $-64a^3 b^6$ **12.** $81a^{24} b^4 c^{16}$ **13.** $-14b^9$ **14.** $\frac{1}{6}y^{11}$ **15.** $12a^7 b^4$ **16.** $\frac{20}{3}a^3 b^2$

Section 5.3

Five-Minute Warm-Up **1.** x^7 **2.** $20y^8$ **3.** $64z^6$ **4.** $9a^8$ **5.** $\frac{25}{4}x^2$ **6.** $-16x^{11} y^{16}$ **7.** $44a - 33b$ **8.** $\frac{4x}{5} + \frac{3}{4}$

Guided Practice **1.** the Distributive Property **2a.** $-15a^3 b^2 + 6a^2 b^3 - 9ab^4$ **2b.** $42x^7 - 3x^6 - 8x^5$ **3a.** $x^2 - 6x + 8$ **3b.** $6x^2 - 7x - 5$ **4a.** multiply two binomials **4b.** First, Outer, Inner, Last **4c.** no **5a.** $u^2 - 2u - 15$ **5b.** $7x^2 - 23xy + 6y^2$ **6.** $A^2 - B^2$ **7a.** $x^2 - 81$ **7b.** $4a^2 - 25b^2$ **8a.** $A^2 + 2AB + B^2$ **8b.** $A^2 - 2AB + B^2$ **9a.** $(A - B)^2 = A^2 - 2AB + B^2$ **9b.** x **9c.** 3 **9d.** x^2 **9e.** $6x$ **9f.** 9 **9g.** $x^2 - 6x + 9$ **10a.** $144 - 24x + x^2$ **10b.** $16x^2 + 40xy + 25y^2$ **11a.** $3x^3 + 5x^2 - 14x + 4$ **11b.** $8x^3 - 14x^2 - 27x - 9$

Do the Math **1.** $6m^2 - 21m$ **2.** $9b^2 - 3b$ **3.** $8w^3 + 12w^2 - 20w$ **4.** $14r^3 s + 6r^2 s^2$ **5.** $n^2 - 3n - 28$ **6.** $10n^2 - 19ny + 6y^2$ **7.** $x^2 + 10x + 21$ **8.** $12z^2 - 5z - 2$ **9.** $x^4 - 7x^2 + 10$ **10.** $18r^2 + 51rs + 35s^2$ **11.** $36r^2 - 1$ **12.** $64a^2 - 25b^2$ **13.** $x^2 + 8x + 16$ **14.** $36b^2 - 60b + 25$ **15.** $9x^2 + 12xy + 4y^2$ **16.** $y^2 - \frac{2}{3}y + \frac{1}{9}$ **17.** $6a^3 - 17a^2 - 4a + 3$ **18.** $-4k^3 - 16k^2 + 84k$ **19.** $-2m^5 + m^4 - 8m^3 - 7m - 4$ **20.** $2a^3 + 9a^2 + 3a - 4$

Sullivan/Struve/Mazzarella, *Elementary Algebra*, 2e

Section 5.4

Five-Minute Warm-Up **1.** $49x^6y^2$ **2.** $\dfrac{36}{25}$ **3a.** $-\dfrac{1}{3}$ **3b.** $\dfrac{8}{7}$ **4.** $\dfrac{9}{5}$ **5.** $-\dfrac{11}{24}$

Guided Practice **1.** subtract **2a.** x^7 **2b.** $-\dfrac{9}{4}a^2b$ **3.** To raise a quotient to a power, both the numerator and the denominator are raised to the indicated power. The answer is (c) it does not matter which occurs first, although generally it is easier to simplify first. Be sure variables are non-zero. **4a.** $-\dfrac{v^3}{8}$ **4b.** $\dfrac{49a^{10}}{16b^8}$ **5.** 1

6a. 1 **6b.** 1 **6c.** 12 **6d.** −1 **6e.** 1 **6f.** −4 **7a.** $\dfrac{1}{a^n}$ **7b.** a^n **7c.** $\left(\dfrac{b}{a}\right)^n = \dfrac{b^n}{a^n}$ **8a.** $-\dfrac{1}{32}$ **8b.** $-\dfrac{1}{12}$ **8c.** $\dfrac{5}{a^2}$

9a. $\left(\dfrac{4}{3}\cdot(-18)\right)(x^{-2}\cdot x^5)(y^4\cdot y^{-7})$ **9b.** −24 **9c.** x^3 **9d.** y^{-3} **9e.** $-\dfrac{24x^3}{y^3}$

Do the Math **1.** 1000 **2.** x^6 **3.** $\dfrac{-3xy}{2}$ **4.** $3rs$ **5.** $\dfrac{16}{81}$ **6.** $\dfrac{49}{x^4}$ **7.** $-\dfrac{a^{15}}{b^{50}}$ **8.** $\dfrac{16m^4n^8}{q^{12}}$ **9.** −1 **10.** 1 **11.** 1

12. $-11x$ **13.** $-\dfrac{7}{z^5}$ **14.** $\dfrac{p^4}{16}$ **15.** $4b^3$ **16.** $36t$ **17.** 1 **18.** $\dfrac{2a^2}{b^2}$ **19.** $\dfrac{x^8}{25y^6}$ **20.** $20a^5$

Section 5.5

Five-Minute Warm-Up **1.** $-27x^4$ **2.** $\dfrac{1}{9n^3}$ **3.** $-27z^3 - 18z^2 + 9z$ **4a.** 3 **4b.** $-7x^3 - 3x^2 + 5x + 13$
5. −1 **6.** $-x - 2$

Guided Practice **1a.** $2x^2 + 9x$ **1b.** $x + 4y^2 - 1$ **2.** Multiply the quotient by the divisor. Add the remainder to this product. The result will be equal to the dividend if the problem was divided correctly.

3a. 71 **3b.** 8047 **3c.** 113 **3d.** 24 **3e.** $113\dfrac{24}{71}$ **4a.** x^2 **4b.** $x^3 + x^2$ **4c.** $-3x^2 + x + 6$

4d. $x^2 - 3x + 4 + \dfrac{2}{x+1}$ **5.** standard **6.** False

Do the Math **1.** $x - 2$ **2.** $2m + 1 - \dfrac{1}{2m^2}$ **3.** $\dfrac{4}{5}a - \dfrac{3}{5} + \dfrac{2}{5a}$ **4.** $-\dfrac{7}{6} - \dfrac{7}{2p} + \dfrac{1}{2p^2}$ **5.** $-1 + 3y^2 - \dfrac{16y^3}{5}$

6. $-7x - 4y$ **7.** $-\dfrac{3y^2}{x^2} - 5$ **8.** $\dfrac{2}{mn} + \dfrac{3m}{n}$ **9.** $x + 8$ **10.** $x^2 - 7x + 2$ **11.** $x^3 - x^2 + 1$ **12.** $x^2 - 4x + 3 - \dfrac{2}{x-3}$

13. $3x - 2 - \dfrac{10}{2+3x}$ **14.** $16x^4 + 12x^2 + 9$

Section 5.6

Five-Minute Warm-Up **1.** $19x^5$ **2.** $2.56n^{-6}$ or $\dfrac{2.56}{n^6}$ **3.** $0.2p^4$ **4a.** 4200 **4b.** 0.00305 **4c.** 2.1 **5a.** 10^8

5b. 10^{-9} **5c.** 10^5 **5d.** 10^{-8} **5e.** 10^5
Guided Practice **1a.** $1 \le x < 10$ **1b.** any integer **1c.** positive **1d.** negative **2a.** 7 **2b.** positive
2c. 4.5×10^7 **3a.** 5 **3b.** negative **3c.** 3×10^{-5} **4a.** right **4b.** left **5a.** 4 **5b.** right **5c.** 60,200 **6a.** −3
6b. left **6c.** 0.0091 **7a.** 8×10^{13} **7b.** 5.4×10^{-4} **8a.** 4×10^5 **8b.** 9×10^{36}

Do the Math **1.** 8×10^9 **2.** 1×10^{-7} **3.** 2.83×10^{-5} **4.** 4.01×10^8 **5.** 8×10^0 **6.** 1.2×10^2 **7.** 375
8. 6,000,000 **9.** 0.0005 **10.** 0.49 **11.** 540,000 **12.** 0.005123 **13.** 2.4×10^{-8} **14.** 1×10^4 **15.** 5×10^{-2}
16. 4×10^5 **17.** 1.1×10^{-7} **18.** 2×10^{11}

Chapter 6 Answers

Section 6.1

Five-Minute Warm-Up **1a.** $2 \bullet 2 \bullet 3 \bullet 3$ **1b.** $3 \bullet 3 \bullet 5 \bullet 5$ **2.** $-12x + 27$ **3.** $2x^2 + 5x - 12$

4. $4x^4 - 36x^2$ **5a.** $\dfrac{1}{2}, 10$ **5b.** 5 **6.** $4x^2y$

Guided Practice **1a.** $3 \bullet 3 \bullet 3$ **1b.** $2 \bullet 2 \bullet 3 \bullet 3$ **1c.** $3 \bullet 3 \bullet 5$ **1d.** $3 \bullet 3$ **1e.** 9 **2a.** $6x^2$ **2b.** $3ab$

3a. $2m^2n^2$ **3b.** $2m^2n^2 \bullet 3m^2 + 2m^2n^2 \bullet 9mn^2 - 2m^2n^2 \bullet 11n^3$ **3c.** $2m^2n^2\left(3m^2 + 9mn^2 - 11n^3\right)$

4a. $-3x^3\left(8x^2 + 3\right)$ **4b.** $(x + 3)(4x + 7)$ **5.** 4 **6.** True **7a.** $2x$ **7b.** $3y$ **7c.** $2x(x - 2) + 3y(x - 2)$

7d. $(x - 2)(2x + 3y)$ **8.** $(x - 2y)(9 - 4a)$ **9.** $2(2x + 3)\left(x^2 - 3\right)$

Do the Math **1.** 7 **2.** $13xy$ **3.** xyz **4.** $x + y$ **5.** $3(a - b)$ **6.** $-3b(3b + 2a)$ **7.** $4a^3b^2\left(2 + 3a^2\right)$

8. $5x^2(x^2 + 2x - 5)$ **9.** $-2(y^2 - 5y + 7)$ **10.** $-2n(11n^3 - 9n - 7)$ **11.** $(a - 5)(a + 6)$ **12.** $(x + a)(x + 2)$

13. $(m - 3)(n + 2)$ **14.** $(z + 4)\left(z^2 + 3\right)$ **15.** $(x - 1)\left(x^2 - 5\right)$

Section 6.2

Five-Minute Warm-Up **1.a.** $-1, 12; -2, 6; -3, 4; 1, -12; 2, -6; 3, -4$ **1b.** $1, 36; 2, 18; 3, 12; 4, 9;$

1b. $-1, -36; -2, -18; -3, -12; -4, -9$ **2a.** -15 **2b.** 36 **3a.** 6 **3b.** -16 **4a.** $4, -9$ **4b.** $-3, -2$

4c. $-5, 1$ **4d.** $3, 6$ **5.** $-1, 7, -2$ **6.** $x^2 - 10x + 24$

Guided Practice **1a.** $19; 11; 9; -19; -11; -9$ **1b.** $2, 9$ **1c.** $(y + 2)(y + 9)$

2a. $37; 20; 15; 13; -37; -20; -15; -13$ **2b.** $(z - 3)(z - 12)$ **3a.** See Graphing Answer Section

3b. $(x - 7)(x + 8)$ **4a.** See Graphing Answer Section **4b.** $(n + 4)(n - 6)$ **5.** Factors of -6 are:
$-1, 6; 1, -6; -2, 3; 2, -3$. The sums are $5, -5, 1, -1$. Since there are no two factors of -6 whose sum is 3, the polynomial is prime. **6.** $3p(p - 6)(p + 3)$

Do the Math **1.** $(n + 2)(n + 10)$ **2.** $(y + 1)(y - 9)$ **3.** $(y - 4)(y + 10)$ **4.** prime **5.** $(x - 2y)(x - 12y)$

6. $4p^2(p - 2)(p + 1)$ **7.** $-2(x - 10)(x - 5)$ **8.** $-3(x + 1)(x + 5)$ **9.** $x(x + 15)(x - 5)$ **10.** $-2z(z + 4)(z - 3)$

11. $-(r + 6)^2$ **12.** prime **13.** $x(x - 8)(x + 2)$ **14.** $(x + 3y)(x + 4y)$ **15.** $(x + 5)^2$

Section 6.3

Five-Minute Warm-Up **1.** $4, 1, -1$ **2a.** $2 \bullet 3 \bullet 3$ **2b.** $2 \bullet 3 \bullet 5 \bullet 5$ **3.** $6x^2 - x - 2$ **4.** $(x - y)(4 + a)$

5. $3rs\left(4r^2s + 1 - 2s^3\right)$ **6.** $(2z + 1)(7z - 4)$

Guided Practice **1a.** $(3x + ?)(x + ?)$ **1b.** $1, 2; -1, -2$ **1c.** $1, 2$ **1d.** $(3x + 1)(x + 2)$

2. $(2a - 3b)(9a - 4b)$ **3.** $-(3x + 2)(2x - 7)$ **4a.** $3; 12$ **4b.** 36 **4c.** $-37; -20; -15; -13; -12$ **4d.** $-4, -9$

4e. $3x^2 - 4x - 9x + 12 = x(3x - 4) - 3(3x - 4)$ **4f.** $(x - 3)(3x - 4)$ **5.** $(13x - 4)(x + 1)$

Do the Math **1.** $(3x - 2)(x + 6)$ **2.** $(5x + 1)(x + 3)$ **3.** $(11p - 2)(p - 4)$ **4.** $(3x + y)(x + 2y)$

5. $2(3x + 2y)(x - 3y)$ **6.** $-3(2x - 5)(x + 3)$ **7.** $(7n + 1)(n - 4)$ **8.** $(5t - 1)(5t + 2)$ **9.** $(5t + 4)(4t + 1)$

10. $(4p - 1)(3p - 5)$ **11.** $2(3x + 2y)(3x - y)$ **12.** $-(10y + 3)(y - 5)$ **13.** $2(x + 5)(9x - 1)$ **14.** $3xy(3x^2 + 2x + 1)$

15. prime

Section 6.4

Five-Minute Warm-Up **1.** 100 **2.** -27 **3.** $64x^6$ **4.** $4p^2 + 20p + 25$ **5.** $16a^2 - 81b^2$

6. $1, 4, 9, 16, 25, 36, 49, 64, 81$ **7.** $1, 8, 27, 64$

Guided Practice **1a.** $A^2 + 2AB + B^2$ **1b.** $A^2 - 2AB + B^2$ **2a.** z^2, z **2b.** $16; 4$ **2c.** $-8z$, yes,
$-8z = -2(z)(4)$ **2d.** $(z - 4)^2$ **3a.** $(p + 11)^2$ **3b.** $3n(2m - 3n)^2$ **4.** $A^2 - B^2$ **5a.** $(x - 7)(x + 7)$

5b. $\left(5m^3 - 6n^2\right)\left(5m^3 + 6n^2\right)$ **5c.** $(a - 2)(a + 2)\left(a^2 + 4\right)\left(a^4 + 16\right)$ **5d.** $3y(2x - 5y)(2x + 5y)$

6a. $A^3 + B^3$ **6b.** $A^3 - B^3$ **6c.** $A^3 + 3A^2B + 3AB^2 + B^3$ **7a.** $(p - 4)\left(p^2 + 4p + 64\right)$

7b. $\left(2x^2 + 3y\right)\left(4x^4 - 6x^2y + 9y^2\right)$ **8a.** $3x(2x + 5)\left(4x^2 - 10x + 25\right)$

8b. $2xy(5x - 2y)\left(25x^2 + 10xy + 4y^2\right)$

Sullivan/Struve/Mazzarella, *Elementary Algebra*, 2e

Do the Math **1.** $(m+6)^2$ **2.** $(3a-2)^2$ **3.** $(4y-9)^2$ **4.** $(2a+5b)^2$ **5.** $(6m-5n)(6m+5n)$

6. $(a-2)(a+2)(a^2+4)$ **7.** $(4r-5s)(16r^2+20rs+25s^2)$ **8.** $(m^3-3n^2)(m^6+3m^3n^2+9n^4)$

9. $(5y+3z^2)(25y^2-15yz^2+9z^4)$ **10.** $5(2x+3y^2)(4x^2-6xy^2+9y^4)$ **11.** $3n(2n-3)^2$

12. $4(2a+b^2)(4a^2-2ab+b^4)$ **13.** $x^2(x-15)(x+15)$ **14.** prime **15.** $3(x+3)^2$

Section 6.5

Five-Minute Warm-Up **1.** $20x^2+7xy-6y^2$ **2.** $a^2+8ab+16b^2$ **3.** $-9a(3a^2-a+2)$

4. $3(2n-1)(2m-3)$ **5.** $(p-2q)(4p+3)$

Guided Practice **1.** factor out the GCF **2a.** 4 **2b.** $4(x^2+4x-21)$ **2c.** 3 **2d.** $4(x+7)(x-3)$ **3a.** 2

3b. $(8a-9b)(8a+9b)$ **4a.** 2 **4b.** $2(4a^2+12ab+9b^2)$ **4c.** 3 **4d.** $4a^2$; $2a$ **4e.** $9b^2$; $3b$

4f. $12ab$; yes, $12ab=2(2a)(3b)$ **4g.** $2(2a+3b)^2$ **5.** $(3p^2+1)(9p^4-3p^2+1)$ **6.** $2(n^2-3)(n-5)$

7. $-3x(y+1)(y+3)$

Do the Math **1.** $(x-16)(x+16)$ **2.** $(1-y)(1+y+y^2)(1+y^3+y^6)$ **3.** $(x+6)(x-1)$

4. $(2x-1)(x-3)(x+3)$ **5.** $25(2x-y)(2x+y)$ **6.** $(3st-1)(2st+1)$ **7.** $(2x^2+5y)(4x^4-10x^2y+25y^2)$

8. $3m(2-m^2)(2+m^2)(4+m^4)$ **9.** $4t^2(t^2+4)$ **10.** $2n(2n-3)(2n+3)$ **11.** prime

12. $3q(2p+3q)(4p^2-6pq+9q^2)$ **13.** $(x+y)(x-4)(x+4)$ **14.** $-3x(x+3)(x+1)$ **15.** $a(2a-b)(3a-b)$

Section 6.6

Five-Minute Warm-Up **1.** $\left\{\dfrac{1}{2}\right\}$ **2.** $\{7\}$ **3a.** 15 **3b.** 10 **4.** $(x+9)(x-7)$ **5.** $(2p+3)(2p-1)$

6. $(x-9)(x+9)$ **7.** $(n-8)^2$

Guided Practice **1.** If the product of two numbers is zero, then one of the factors must be zero. That is, if

$a \bullet b = 0$, then $a=0$ or $b=0$ or both a and b are zero. **2.** $\left\{-1, \dfrac{3}{2}\right\}$ **3.** quadratic equation

4. List the terms from the highest power to the lowest power. **5a.** $(x+8)(x-3)=0$ **5b.** $x+8=0$

5c. $x-3=0$ **5d.** $x=-8$ **5e.** $x=3$ **5f.** $\{-8, 3\}$ **6a.** $3t^2+11t-4=0$ **6b.** $(3t-1)(t+4)=0$

6c. $\left\{\dfrac{1}{3}, -4\right\}$ **7a.** multiply the binomials **7b.** standard form **7c.** $(x-10)(x+5)=0$ **7d.** $\{10, -5\}$

8a. $4x^2-24x+36=0$ **8b.** $4(x-3)^2=0$ **8c.** $\{3\}$ **9a.** no **9b.** 4 **9c.** factor by grouping

9d. $(p+2)(p-3)(p+3)=0$ **9e.** $p+2=0$ **9f.** $p-3=0$ **9g.** $p+3=0$ **9h.** $p=-2$ **9i.** $p=3$

9j. $p=-3$ **9k.** $\{2, 3, -3\}$

Do the Math **1.** $\{-9, 0\}$ **2.** $\left\{-4, \dfrac{3}{4}\right\}$ **3.** linear **4.** quadratic **5.** $\{-3, 8\}$ **6.** $\left\{0, \dfrac{2}{7}\right\}$ **7.** $\left\{-\dfrac{7}{3}, 2\right\}$ **8.** $\{-6\}$

9. $\{-2, 8\}$ **10.** $\{-3, 10\}$ **11.** $\left\{-\dfrac{3}{2}, \dfrac{1}{2}\right\}$ **12.** $\{-6, 4\}$ **13.** $\left\{-\dfrac{7}{3}, 0, 2\right\}$ **14.** $\{-3, -2, 3\}$ **15.** 1 sec, 3 sec

Section 6.7

Five-Minute Warm-Up **1.** $\{-9, 5\}$ **2.** $\{-6\}$ **3.** $\left\{-\dfrac{5}{2}, -\dfrac{1}{3}\right\}$ **4.** 15 in. **5.** $x^2-10x+25$

6. $-9h(2h-1)(h-1)$

Guided Practice **1a.** 0 sec or 3 sec **1b.** 4 sec **2a.** $n+2$ **2b.** $A = l \cdot w$ **2c.** $n; n+2$

2d. $255 = n(n+2)$ **2e.** $n = 15$ or $n = -17$ **2f.** The rectangle has dimensions $15\ cm \times 17\ cm$ **3.** In a right triangle, the square of the length of the hypotenuse is equal to the sum of the squares of the lengths of the legs. **4.** $z^2 = x^2 + y^2$ **5a.** $30\ cm$ **5b.** $7\ ft$ **6a.** $x = 0$ or $x = 9$ **6b.** leg: $9\ m$; leg: $12\ m$; hypotenuse: $15\ m$

Do the Math **1.** 10, 25 **2.** 14, 26 **3.** base = 6; height = 14 **4.** base = 12; height = 14 **5.** base = 11; height = 7 **6.** $B = 17; h = 5; b = 11$ **7.** 9, 12, 15 **8.** 15, 8, 17 **9.** 3 sec **10.** width = 3 ft; length = 7 ft **11.** base = 8m; height = 6m **12.** length = 13 yd; width = 8yd **13a.** width = 26 in.; height = 36 in. **13b.** No

Chapter 7 Answers
Section 7.1

Five-Minute Warm-Up **1.** -2 **2.** $(4x-3)(x+1)$ **3.** $\left\{-\dfrac{5}{2}, 3\right\}$ **4.** $\dfrac{10}{21}$ **5.** $\dfrac{2x}{3y^2}$

Guided Practice **1a.** $\dfrac{3}{10}$ **1b.** 4 **2a.** $-\dfrac{2}{9}$ **2b.** $\dfrac{7}{4}$ **3.** the denominator is equal to zero. **4a.** $x = -8$

4b. $x = 3$ or $x = 6$ **5a.** factor **5b.** divide out **6a.** $\dfrac{6(x-3)}{(x-2)(x-3)}$ **6b.** $\dfrac{6\cancel{(x-3)}}{(x-2)\cancel{(x-3)}}$ **6c.** $\dfrac{6}{x-2}$

7. $\dfrac{x+1}{2(x+2)}$ **8.** $-\dfrac{x-5}{2(x+2)}$

Do the Math **1a** $\dfrac{2}{3}$ **1b.** undefined **1c.** 0 **2a.** $\dfrac{7}{4}$ **2b.** 1 **2c.** 3 **3a.** 15 **3b.** $-\dfrac{3}{5}$ **3c.** -3 **4a.** -5 **4b.** -9 **4c.** 0

5. -8 **6.** $\dfrac{3}{4}$ **7.** $-2, 1$ **8.** $-4, -1, 0$ **9.** $2p+1$ **10.** -1 **11.** $\dfrac{x-3}{x+2}$ **12.** $\dfrac{x+2}{x-3}$ **13.** $-\dfrac{1}{b+a}$ **14.** $\dfrac{z^2+2}{z+2}$

15. $\dfrac{x+3}{2x-3}$ **16.** $-\dfrac{x+1}{x+5}$

Section 7.2

Five-Minute Warm-Up **1.** $\dfrac{27}{14}$ **2.** $-\dfrac{2}{5}$ **3.** $\dfrac{1}{7}$ **4.** $-5x(x-5)$ **5.** $\dfrac{x-3}{x-6}$

Guided Practice **1a.** $\dfrac{x+3}{8} \cdot \dfrac{4(x-3)}{(x+3)(x-3)}$ **1b.** $\dfrac{4(x+3)(x-3)}{8(x+3)(x-3)}$ **1c.** $\dfrac{1\cancel{4}\cancel{(x+3)}\cancel{(x-3)}}{2\cancel{8}\cancel{(x+3)}\cancel{(x-3)}}$ **1d.** $\dfrac{1}{2}$

2a. $\dfrac{x(x-4)}{(x-2)(x+2)} \cdot \dfrac{(x-3)(x+2)}{(x-4)}$ **2b.** $\dfrac{x(x-4)(x-3)(x+2)}{(x-2)(x+2)(x-4)}$ **2c.** $\dfrac{x\cancel{(x-4)}(x-3)\cancel{(x+2)}}{(x-2)\cancel{(x+2)}\cancel{(x-4)}}$

2d. $\dfrac{x(x-3)}{(x-2)}$ **3.** $-\dfrac{2(x-y)}{x(x+y)}$ **4a.** $\dfrac{3x+15}{36} \cdot \dfrac{24}{5x+25}$ **4b.** $\dfrac{3(x+5)}{2\cdot2\cdot3\cdot3} \cdot \dfrac{2\cdot2\cdot2\cdot3}{5(x+5)}$

4c. $\dfrac{3\cdot(x+5)\cdot2\cdot2\cdot2\cdot3}{3\cdot2\cdot2\cdot3\cdot5\cdot(x+5)}$ **4d.** $\dfrac{\cancel{3}\cdot\cancel{(x+5)}\cdot\cancel{2}\cdot\cancel{2}\cdot2\cdot\cancel{3}}{\cancel{3}\cdot\cancel{2}\cdot\cancel{2}\cdot\cancel{3}\cdot5\cdot\cancel{(x+5)}}$ **4e.** $\dfrac{2}{5}$ **5.** $\dfrac{1}{2x-3}$ **6.** $-\dfrac{3}{2x}$

Do the Math **1.** $\dfrac{3x}{x+4}$ **2.** $\dfrac{2(n+2)}{3(n-2)}$ **3.** $z+2$ **4.** $\dfrac{x+2}{x-2}$ **5.** $\dfrac{2}{p}$ **6.** $\dfrac{z+5}{2(z-5)}$ **7.** $\dfrac{9y(y-1)}{2}$ **8.** $\dfrac{3}{2}$

9. $\dfrac{1}{(2x-3)(x+4)}$ **10.** $\dfrac{2(x+5)^2}{45x}$ **11.** $r-3s$ **12.** $\dfrac{x^2+1}{(x+1)^2}$ **13.** $\dfrac{2x^3(x-3)}{(x+3)}$ **14.** $-\dfrac{(p-7)}{(p+2)(p-2)}$ **15.** $\dfrac{b-9}{b-3}$

16. $\dfrac{t(t+3)}{(t+4)(t-3)}$

Sullivan/Struve/Mazzarella, *Elementary Algebra*, 2e

Section 7.3

Five-Minute Warm-Up **1.** $\dfrac{3}{5}$ **2a.** $\dfrac{13}{3}$ **2b.** $\dfrac{5}{2}$ **3a.** 1 **3b.** $\dfrac{5}{8}$ **4.** $-\dfrac{1}{2}$ **5.** $-2x+3$

Guided Practice **1a.** $\dfrac{2x+5x}{x^2-3x}=\dfrac{7x}{x^2-3x}$ **1b.** $\dfrac{7x}{x(x-3)}$ **1c.** $\dfrac{7\cancel{x}}{\cancel{x}(x-3)}$ **1d.** $\dfrac{7}{x-3}$ **2.** $\dfrac{5}{x+5}$

3a. $\dfrac{x^2-1}{x^2-5x-6}$ **3b.** $\dfrac{(x-1)(x+1)}{(x-6)(x+1)}$ **3c.** $\dfrac{(x-1)\cancel{(x+1)}}{(x-6)\cancel{(x+1)}}$ **3d.** $\dfrac{x-1}{x-6}$ **4.** $x+2$ **5.** $\dfrac{(n+5)(n-1)}{(n-5)}$

6. $\dfrac{x+1}{x-3}$

Do the Math **1.** $2n$ **2.** $\dfrac{5p}{p-1}$ **3.** $2(x+2)$ **4.** $\dfrac{1}{x-3}$ **5.** a **6.** $\dfrac{2(x+2)}{x}$ **7.** $\dfrac{z+6}{z-1}$ **8.** $\dfrac{n+3}{n+1}$ **9.** $\dfrac{4(1-2x)}{x-1}$

10. $\dfrac{x^2+x+6}{x-1}$ **11.** $\dfrac{x+3}{x+1}$ **12.** $\dfrac{m+n}{n-m}$ **13.** $\dfrac{3n(n-2)}{2n-1}$ **14.** $\dfrac{x+3}{x-3}$ **15.** $\dfrac{4s-t}{t-s}$ **16.** $\dfrac{n+3}{2n-3}$

Section 7.4

Five-Minute Warm-Up **1.** $\dfrac{21}{45}$ **2.** 180 **3.** $\dfrac{9}{20}=\dfrac{81}{180};\dfrac{2}{45}=\dfrac{8}{180}$ **4.** $3y^2$ **5.** $(x-2)^2$ **6.** $-3x(x+3)$

Guided Practice **1a.** $2\bullet2\bullet2\bullet3$ **1b.** $2\bullet2\bullet5$ **1c.** $2\bullet2$ **1d.** $2\bullet3\bullet5$ **1e.** 120 **2a.** $2\bullet3\bullet x^2$

2b. $3^2\bullet x$ **2c.** $3\bullet x$ **2d.** $2\bullet3\bullet x$ **2e.** $18x^2$ **3.** $40a^3b^2$ **4.** $14a^2(a-2)$ **5.** $(x-3)^2(x+3)$ **6a.** $x(3x-1)$

6b. $3x^2(3x-1)$ **6c.** $3x$ **6d.** $\dfrac{3x}{3x}$ **6e.** $\dfrac{x-2}{x(3x-1)}\bullet\left[\dfrac{3x}{3x}\right]=\dfrac{3x^2-6x}{3x^2(3x-1)}$

7. $\text{LCD}=36x^2y^2;\dfrac{4}{9xy^2}=\dfrac{16x}{36x^2y^2};\dfrac{5}{12x^2}=\dfrac{15y^2}{36x^2y^2}$

Do the Math **1.** $49y^2$ **2.** $24(x-3)$ **3.** $4(x-3)^2(x+3)$ **4.** $(x-1)^2(x+1)$ **5.** $-3(c-7)(c+7)$

6. $-(z-4)(z+5)$ **7.** $\dfrac{7a^2b+ab}{a^2b^2c}$ **8.** $\dfrac{3x+9}{(x+2)(x+3)}$ **9.** $\dfrac{14a}{6(a+1)(a+2)}$ **10.** $\dfrac{7t^2+7}{t^2+1}$ **11.** $\dfrac{28}{35n^2};\dfrac{15n}{35n^2}$

12. $\dfrac{8p^3-4p}{24p^2};\dfrac{9p^3+6}{24p^2}$ **13.** $\dfrac{(x+2)^2}{x(x+2)};\dfrac{x^2}{x(x+2)}$ **14.** $\dfrac{3ab+6b^2}{7a(a+2b)};\dfrac{2a^2}{7a(a+2b)}$ **15.** $\dfrac{5}{m-6};\dfrac{-2m}{m-6}$

16. $\dfrac{4a}{(a+1)(a-1)};\dfrac{7a-7}{(a+1)(a-1)}$ **17.** $\dfrac{4n^2+8n}{(n+2)^2(n-3)};\dfrac{2n-6}{(n+2)^2(n-3)}$ **18.** $\dfrac{3}{(n-3)(2n-1)};\dfrac{4n^2-2n}{(n-3)(2n-1)}$

Section 7.5

Five-Minute Warm-Up **1a.** $\text{LCD}=6;\dfrac{3}{2}=\dfrac{9}{6};\dfrac{5}{3}=\dfrac{10}{6}$ **1b.** $\text{LCD}=84;\dfrac{7}{12}=\dfrac{49}{84};\dfrac{5}{28}=\dfrac{15}{84}$

2. $(3z-2)(2z-1)$ **3.** $2x-1$ **4.** $-\dfrac{x+4}{2x}$ **5.** $\dfrac{1}{2x}$

Guided Practice **1a.** $3\bullet5$ **1b.** 5^2 **1c.** $3\bullet5^2=75$ **1d.** $\dfrac{7}{15}\bullet\left[\dfrac{5}{5}\right]=\dfrac{35}{75}$ **1e.** $\dfrac{10}{25}\bullet\left[\dfrac{3}{3}\right]=\dfrac{30}{75}$

1f. $\dfrac{35+30}{75}=\dfrac{65}{75}$ **1g.** $\dfrac{13}{15}$ **2a.** $2^2\bullet3\bullet x$ **2b.** $2\bullet3^2\bullet x^2$ **2c.** $2^2\bullet3^2\bullet x^2=36x^2$ **2d.** $\dfrac{5}{12x}\bullet\left[\dfrac{3x}{3x}\right]=\dfrac{15x}{36x^2}$

2e. $\dfrac{7x}{18x^2}\bullet\left[\dfrac{2}{2}\right]=\dfrac{14x}{36x^2}$ **2f.** $\dfrac{15x+14x}{36x^2}=\dfrac{29x}{36x^2}$ **2g.** $\dfrac{29}{36x}$ **3.** $\dfrac{6x+8}{(x+2)(x+1)}$ **4a.** $x(x-2)$

4b. $\dfrac{2}{x}\bullet\left[\dfrac{x-2}{x-2}\right]=\dfrac{2x-4}{x(x-2)}$ **4c.** $\dfrac{4}{x-2}\bullet\left[\dfrac{x}{x}\right]=\dfrac{4x}{x(x-2)}$ **4d.** $\dfrac{2x-4-4x}{x(x-2)}=\dfrac{-2x-4}{x(x-2)}$ **4e.** $\dfrac{-2(x+2)}{x(x-2)}$

5. 1 **6.** $\dfrac{2x-10}{x-4}$

Do the Math 1. $\dfrac{4}{3}$ **2.** $\dfrac{25+12x}{10x}$ **3.** $\dfrac{x^2+3}{(x-1)(x+1)}$ **4.** $-\dfrac{1}{n-4}$ **5.** $\dfrac{x+6}{x+2}$ **6.** $\dfrac{4}{x-4}$ **7.** $\dfrac{22}{105}$ **8.** $\dfrac{-8x}{(x+2)(x-2)}$

9. $\left(\dfrac{x+2}{x+4}\right)^2$ **10.** $\dfrac{2p^2+4p+1}{p(p-2)}$ **11.** $\dfrac{5n-12}{4n^2}$ **12.** $\dfrac{3x+4}{x+3}$ **13.** $\dfrac{4}{3n(n-3)}$ **14.** $\dfrac{a^2-5a+3}{2a^2(a-3)}$ **15.** $\dfrac{x^2-2x-11}{(x-3)(x-2)}$

16. $\dfrac{15}{w(w-3)}$

Section 7.6

Five-Minute Warm-Up 1. $(2x+3)(x-4)$ **2.** $3(x+5)(x-5)$ **3.** $\dfrac{4}{3(x+2)}$ **4.** x^2-2x+1

5. $3x^2+2x$

Guided Practice 1. When sums and/or differences of rational expressions occur in the numerator or denominator of a quotient, the quotient is called a complex rational expression. Rational expressions with more than one fraction bar are complex and need to be simplified. **2a.** $5x$ **2b.** $\dfrac{25-x^2}{5x}$ **2c.** $5x^2$

2d. $\dfrac{x^2-25}{5x^2}$ **2e.** $\dfrac{\dfrac{25-x^2}{5x}}{\dfrac{x^2-25}{5x^2}}$ **2f.** $\dfrac{25-x^2}{5x}\cdot\dfrac{5x^2}{x^2-25}$ **2g.** $\dfrac{^{(-1)}\cancel{(25-x^2)}\cdot\cancel{5}\cdot\cancel{x}\cdot x}{\cancel{5}\cdot\cancel{x}\cdot\cancel{(25-x^2)}\cdot 1}=\dfrac{-x}{1}=-x$

3a. Multiplicative Inverse (a number multiplied by its reciprocal is one) or alternately, any quotient with the same numerator and denominator is a representation of one whole. **3b.** Multiplicative Identity

3c. Distributive Property **4a.** x^2 **4b.** $\left(\dfrac{1+\dfrac{1}{x}}{1-\dfrac{1}{x^2}}\right)\cdot\left[\dfrac{x^2}{x^2}\right]$ **4c.** $\dfrac{1\bullet x^2+\dfrac{1}{x}\bullet x^2}{1\bullet x^2-\dfrac{1}{x^2}\bullet x^2}$ **4d.** $\dfrac{x^2+x}{x^2-1}$

4e. $\dfrac{x\cancel{(x+1)}}{(x-1)\cancel{(x+1)}}=\dfrac{x}{x-1}$

Do the Math 1. $-t(t-2)$ **2.** $\dfrac{1}{x}$ **3.** $\dfrac{3n+2}{(n-2)(n+1)}$ **4.** $\dfrac{2ab}{(a+b)^2}$ **5.** $\dfrac{5c+2d}{2(5c-4d)}$ **6.** $\dfrac{a+8}{4-a}$ **7.** $\dfrac{x-1}{x+1}$

8. $\dfrac{2xy+1}{3x+1}$ **9.** $\dfrac{x-2}{x-3}$ **10.** $\dfrac{6}{y+5}$ **11.** $\dfrac{2x+1}{x-3}$ **12.** $\dfrac{5m+n}{5m}$ **13.** $\dfrac{y}{x}$ **14.** $\dfrac{4(b-1)}{b}$ **15.** $\dfrac{11(2n+3)}{72}$

Section 7.7

Five-Minute Warm-Up 1. $\left\{\dfrac{8}{3}\right\}$ **2.** $(x-7)(x+4)$ **3.** $\left\{-\dfrac{1}{2},\dfrac{1}{2}\right\}$ **4.** $x=-8$ or $x=2$ **5.** $P=\dfrac{A}{1+rt}$

Guided Practice 1. the denominator being equal to zero. **2a.** $x=7$, $x=-7$ **2b.** $(x-7)(x+7)$

2c. $(x-7)(x+7)\dfrac{x+5}{x-7}=\dfrac{x-3}{x+7}(x-7)(x+7)$ **2d.** $(x+7)(x+5)=(x-3)(x-7)$

2e. $x^2+12x+35=x^2-10x+21$ **2f.** $x=-\dfrac{7}{11}$ **2g.** $\left\{-\dfrac{7}{11}\right\}$ **3a.** $x=0$ **3b.** $10x$ **3c.** $90+8=7x$

3d. $\{14\}$ **4a.** $x=1$, $x=-1$ **4b.** $(x-1)(x+1)$ **4c.** $x=-1$ **4d.** $\varnothing$ **5a.** $y=0$ **5b.** xyz

5c. $(xyz)\left(\dfrac{6}{z}\right)=\dfrac{2}{x}(xyz)+\dfrac{1}{y}(xyz)$ **5d.** $6xy=2yz+xz$ **5e.** $6xy-2yz=xz$ **5f.** $y(6x-2z)=xz$

5g. $y=\dfrac{xz}{6x-2z}$ **5h.** $y=\dfrac{xz}{6x-2z}$

Sullivan/Struve/Mazzarella, *Elementary Algebra*, 2e

Do the Math 1. $\left\{\dfrac{12}{13}\right\}$ 2. $\{36\}$ 3. $\left\{-\dfrac{16}{9}\right\}$ 4. $\{\ \}$ or $\varnothing$ 5. $\left\{-\dfrac{8}{3}\right\}$ 6. $\{7\}$ 7. $\left\{-\dfrac{3}{2},\dfrac{2}{3}\right\}$ 8. $\{\ \}$ or $\varnothing$

9. $\{-3,4\}$ 10. $\{-5\}$ 11. $b=\dfrac{a}{c}-2$ or $b=\dfrac{a-2c}{c}$ 12. $b=\dfrac{4ik}{3k-8i}$ 13. $x=\dfrac{ck-m}{cB}$ 14. $a=\dfrac{bX}{X-b}$

15. after 1 hour and after 3 hours 16. 40 or 125 bicycles

Section 7.8

Five-Minute Warm-Up 1. $\{12\}$ 2. $\left\{\dfrac{2}{3},4\right\}$ 3. $\{3,-1\}$

Guided Practice 1. $\dfrac{3}{x}$ Answers may vary 2. $\dfrac{3}{x}=\dfrac{x+2}{9}$ Answers may vary 3. $\left\{\dfrac{13}{3}\right\}$ 4. about 26.3 million flight hours in 2001. 5. The figures are the same shape, but they are a different size. The corresponding answers are equal and the corresponding sides are proportional. 6a. $\dfrac{6}{3.2}=\dfrac{x}{8}$ 6b. 15 ft

7a. $\dfrac{1}{15}$ 7b. $\dfrac{1}{t}$ 7c. $\dfrac{1}{t+3}$ 8a. $\dfrac{1}{3}$ 8b. $\dfrac{1}{5}$ 8c. $\dfrac{1}{t}$ 8d. $\dfrac{1}{3}+\dfrac{1}{5}=\dfrac{1}{t}$ 8e. It takes 1.875 hours to clean the building when Josh and Ken work together. 9a. $180+w$ 9b. $180-w$ 9c. $\dfrac{1000}{180+w}$ 9d. $\dfrac{600}{180-w}$

9e. $\dfrac{1000}{180+w}=\dfrac{600}{180-w}$ 9f. The speed of the wind is 45 mph.

Do the Math 1. $\left\{\dfrac{20}{9}\right\}$ 2. $\{2\}$ 3. $\{-9\}$ 4. $\{-4,-3\}$ 5. $\{-2,6\}$ 6. $\{-3,6\}$ 7. $\dfrac{n}{2}$ 8. $r+3$ 9. 5 lb

10. 1600 pesos 11. 20 m 12. $\dfrac{28}{11}\approx 2.5$ hr 13. 45 mph 14. 9 mph

Section 7.9

Five-Minute Warm-Up 1a. $\left\{\dfrac{15}{2}\right\}$ 1b. $\{32\}$ 1c. $\left\{\dfrac{2}{3}\right\}$ 1d. $\{6\}$ 2. 15.50 3. $\dfrac{1}{2}$

Guided Practice 1. $y=kx$ 2. constant of proportionality 3a. $y=-\dfrac{2}{3}x$ 3b. $y=-\dfrac{25}{3}$ 3c. $C=46.99$

4a. $p=0.008b$ 4b. \$1200 5. $y=\dfrac{k}{x}$ 6. $y=\dfrac{8}{3}$ 7a. $I=\dfrac{k}{R}$ 7b. $k=9000$ 7c. $I=\dfrac{9000}{R}$; $I=75$ amps

Do the Math 1. $y=\dfrac{1}{3}x$ 2. $y=-\dfrac{5}{12}x$ 3. (a) $d=40t$, (b) $d=200$ 4. (a) $m=\dfrac{4}{3}n$, (b) $m=-\dfrac{32}{15}$

5. $y=\dfrac{70}{x}$ 6. $y=-\dfrac{1}{6x}$ 7. (a) $y=\dfrac{12}{x}$, (b) $y=\dfrac{2}{3}$ 8. (a) $f=\dfrac{5}{3d}$, (b) $d=\dfrac{4}{9}$ 9. $r=64$ 10. $x=84$

11. 25 representatives 12. \$143.50 13. 13.5 meters 14. 28 min

Chapter 8 Answers

Section 8.1

Five-Minute Warm-Up 1. $\{100\}$ 2. $\left\{100,\dfrac{0}{15}\right\}$ 3. $\left\{100,\dfrac{0}{15},\dfrac{8}{-2}\right\}$ 4. $\left\{100,\dfrac{0}{15},\dfrac{8}{-2},-\dfrac{9}{4},7.2\right\}$

5. $\left\{\sqrt{5},\pi\right\}$ 6. $\left\{100,\dfrac{0}{15},\dfrac{8}{-2},-\dfrac{9}{4},7.2,\sqrt{5},\pi\right\}$ 7a. $\dfrac{81}{16}$ 7b. 0.09 7c. 36 7d. -36 8. $\dfrac{3}{5}$

Guided Practice 1a. positive; negative 1b. radical 1c. principal root 1d. radicand 1e. 16; $b^2=a$

1f. not a real number 2a. $2,-2$ 2b. 0 4a. $\dfrac{7}{8}$ 4b. 0.6 4a. 11 4b. -9 5a. 0.2 5b. $-\dfrac{1}{5}$ 6a. 13 6b. 17

7a. rational 7b. irrational 7c. real 8a. irrational 8b. not a real number 8c. rational 9a. $|a|$ 9b. $|n+9|$

Do the Math **1.** $-2, 2$ **2.** $-\dfrac{3}{2}, \dfrac{3}{2}$ **3.** $-7, 7$ **4.** $-0.4, 0.4$ **5.** 13 **6.** 0.3 **7.** -8 **8.** 15 **9.** 21 **10.** 2 **11.** 3.464

12. 4.24 **13.** not a real number **14.** rational; 30 **15.** rational; $\dfrac{7}{8}$ **16.** irrational; 4.90 **17.** $y - 16$ **18.** $|p + q|$

Section 8.2

Five-Minute Warm-Up **1.** $1, 4, 9, 16, 25, 36, 49, 64, 81, 100, 121, 144$ **2.** $1, 8, 27, 64, 125$ **3a.** 9 **3b.** $\dfrac{4}{3}$

3c. not a real number **3d.** -7 **3e.** 0.2 **3f.** 40

Guided Practice **1.** A square root expression is simplified provided that the radicand does not contain factors that are perfect squares. **2.** $\sqrt{a} \cdot \sqrt{b}$ **3a.** 25 **3b.** $\sqrt{25 \bullet 2}$ **3c.** $\sqrt{25} \bullet \sqrt{2} = 5 \bullet \sqrt{2} = 5\sqrt{2}$ **4a.** $2\sqrt{6}$

4b. $-15\sqrt{3}$ **4c.** $-3 + 2\sqrt{3}$ **5a.** $8x^5$ **5b.** $3ab^3\sqrt{5}$ **5c.** $5x^2y^3\sqrt{3y}$ **6.** $\dfrac{\sqrt{a}}{\sqrt{b}}$ **7a.** $\dfrac{\sqrt{5}}{3}$ **7b.** $\dfrac{12a\sqrt{a}}{b^2}$ **7c.** $3x^4$

Do the Math **1.** $3\sqrt{3}$ **2.** $6\sqrt{5}$ **3.** -1 **4.** $2 - \sqrt{3}$ **5.** $5x^4$ **6.** $4p^8\sqrt{2}$ **7.** $9y\sqrt{y}$ **8.** $2n^3\sqrt{15n}$ **9.** $5a^3b^2\sqrt{2b}$

10. $5st^3\sqrt{5}$ **11.** $\dfrac{3}{4}$ **12.** $\dfrac{\sqrt{13}}{2}$ **13.** $\dfrac{x^5}{6}$ **14.** $\dfrac{a^3\sqrt{a}}{b^3}$ **15.** $\dfrac{2ab^3\sqrt{6a}}{c}$ **16.** $\dfrac{b^3}{2}$ **17.** $3xy^4z^4\sqrt{2}$ **18.** $m^3p^3\sqrt{30np}$

19. $\dfrac{-3 + 2\sqrt{3}}{4}$ **20.** $-1 + \sqrt{3}$

Section 8.3

Five-Minute Warm-Up **1a.** Yes **1b.** No **1c.** No **2.** $4x + 14y$ **3.** $3a - 9$ **4.** $36x^3 - 12x^2 + 12x$

5. $(a + b)x$ **6.** $4\sqrt{2}$

Guided Practice **1a.** radicand **1b.** Distributive **2a.** $8\sqrt{10}$ **2b.** $4\sqrt{5} + 2\sqrt{6}$ **2c.** $-4\sqrt{x}$ **3a.** $5\sqrt{2}$

3b. $-2\sqrt{5}$ **3c.** $-8\sqrt{2} - 6\sqrt{3}$ **3d.** $4ab\sqrt{3ab}$ **3e.** $3\sqrt{2x}$ **3f.** $9x\sqrt{3x}$ **3g.** $-\dfrac{11}{4}\sqrt{3}$ **3h.** $-\sqrt{2} + 6\sqrt{5}$

Do the Math **1.** $8\sqrt{2}$ **2.** $2k\sqrt{11k}$ **3.** $-\sqrt{3}$ **4.** $-13\sqrt{30}$ **5.** $-2\sqrt{2}$ **6.** $7\sqrt{2}$ **7.** $5\sqrt{2ab}$ **8.** $26\sqrt{3}$

9. $-5\sqrt{7} - 28$ **10.** $2\sqrt{5y}$ **11.** $22a^2\sqrt{3a}$ **12.** $\dfrac{7}{15}\sqrt{3}$ **13.** $7\sqrt{3p} - 4\sqrt{3}$ **14.** 0 **15.** $5y\sqrt{7x} - 2y\sqrt{5x}$

16. $(9n^6 - 10n^4)\sqrt{6} + 4n^4\sqrt{6n}$ **17.** $2 - 4\sqrt{2}$ **18.** $\dfrac{10}{21}\sqrt{2}$ **19.** $\dfrac{11}{8}\sqrt{3}$ **20.** $\dfrac{7}{3}\sqrt{3}$

Section 8.4

Five-Minute Warm-Up **1.** $64n^8$ **2.** $18x^3 + 6x^2 - 4x$ **3.** $3a^2 + 7ab - 6b^2$ **4.** $4x^2 - 20x + 25$

5. $9x^2 - 49y^2$ **6a.** $5x\sqrt{3x}$ **6b.** $-14\sqrt{5}$

Guided Practice **1a.** $\sqrt{21}$ **1b.** $\sqrt{55mn}$ **2a.** 4 **2b.** $18\sqrt{2}$ **3a.** $5m^3\sqrt{3}$ **3b.** $3p^2\sqrt{6p}$ **4a.** 5

4b. $-48\sqrt{3}$ **5a.** $35 - 5\sqrt{5}$ **5b.** $7\sqrt{3} + 12\sqrt{2}$ **6a.** $19 - 2\sqrt{5}$ **6b.** $58 - 44\sqrt{3}$ **7a.** $A^2 + 2AB + B^2$

7b. $A^2 - 2AB + B^2$ **7c.** $A^2 - B^2$ **8a.** $29 - 12\sqrt{5}$ **8b.** -15

Do the Math **1.** $-5\sqrt{6}$ **2.** $18\sqrt{15}$ **3.** $-10a^4$ **4.** $4q^3\sqrt{3q}$ **5.** $240\sqrt{3}$ **6.** $300n\sqrt{3n}$ **7.** 42 **8.** $6y$ **9.** $4\sqrt{2} + 24$

10. $5\sqrt{5} + 15\sqrt{3}$ **11.** $58 - 15\sqrt{2}$ **12.** $-3 - 2\sqrt{11}$ **13.** $14 - 21\sqrt{2} + 6\sqrt{3} - 9\sqrt{6}$ **14.** $-4x + 15x\sqrt{2}$

15. $11 - 6\sqrt{2}$ **16.** $4x + 12\sqrt{x} + 9$ **17.** -29 **18.** 68

Section 8.5

Five-Minute Warm-Up **1.** $\dfrac{12x}{5y^2}$ **2.** $\dfrac{6x\sqrt{x}}{y^4}$ **3.** 3 **4.** $\dfrac{4\sqrt{3}}{3}$ **5.** 14

Guided Practice 1a. $\dfrac{1}{2}$ **1b.** $\dfrac{\sqrt{7}}{5}$ **2.** The process of rationalizing the denominator of a quotient requires rewriting the quotient, using properties of rational expressions, so that the denominator of the equivalent expression does not contain any radicals. **3.** perfect square **4a.** $\dfrac{\sqrt{3}}{\sqrt{3}}$ **4b.** $\dfrac{\sqrt{5}}{\sqrt{5}}$ **4c.** $\dfrac{\sqrt{2a}}{\sqrt{2a}}$

5a. $\sqrt{3}$ **5b.** $\dfrac{\sqrt{3}}{\sqrt{3}}$ **5c.** $\dfrac{2}{\sqrt{3}} \cdot \dfrac{\sqrt{3}}{\sqrt{3}} = \dfrac{2\sqrt{3}}{3}$ **5d.** $\dfrac{2\sqrt{3}}{3}$ **6a.** $\dfrac{3\sqrt{6}}{8}$ **6b.** $\dfrac{\sqrt{15}}{9}$ **7.** conjugate **8a.** $2+\sqrt{15}; -1$

8b. $3\sqrt{3}-5\sqrt{2}; -23$ **9a.** $2-\sqrt{6}$ **9b.** $\dfrac{2-\sqrt{6}}{2-\sqrt{6}}$ **9c.** $\dfrac{8}{\left(2+\sqrt{6}\right)} \cdot \left[\dfrac{2-\sqrt{6}}{2-\sqrt{6}}\right] = \dfrac{8\left(2-\sqrt{6}\right)}{-2}$

9d. $-4\left(2-\sqrt{6}\right) = -8+4\sqrt{6}$ **10.** $\dfrac{x\sqrt{y}+y\sqrt{x}}{x-y}$

Do the Math **1.** $-\dfrac{1}{3}$ **2.** $4\sqrt{3}$ **3.** $-3x$ **4.** $7a$ **5.** $\dfrac{9\sqrt{11}}{11}$ **6.** $-\dfrac{5\sqrt{3}}{2}$ **7.** $\dfrac{2\sqrt{3p}}{3p^2}$ **8.** $-\dfrac{\sqrt{10ab}}{4b}$ **9.** $\sqrt{10}-3;\ 1$

10. $\sqrt{x}+2\sqrt{y};\ x-4y$ **11.** $\dfrac{8+2\sqrt{10}}{3}$ **12.** $\dfrac{8\sqrt{x}+16}{x-4}$ **13.** $\dfrac{6\sqrt{3}+45}{71}$ **14.** $\dfrac{\sqrt{y}-2y}{1-4y}$ **15.** $3+2\sqrt{2}$ **16.** $\dfrac{\sqrt{6xy}}{2x^2}$

Section 8.6

Five-Minute Warm-Up 1. $\left\{-\dfrac{1}{2}\right\}$ **2.** $\{0, 3, 4\}$ **3.** $16x^2+8x+1$ **4.** $34-24\sqrt{2}$ **5.** $x+4$ **6.** $\{-3, 2\}$

Guided Practice 1a. Yes **1b.** No **2a.** $3x-5=16$ **2b.** $3x=21$ **2c.** $x=7$ **2d.** $\{7\}$ **4a.** $\sqrt{2x+1}=5$
3b. $2x+1=25$ **3c.** $2x=24$ **3d.** $x=12$ **3e.** $\{12\}$ **4.** $\varnothing$ **5a.** $\sqrt{2x-1}=\sqrt{x-1}+1$
5b. $\left(\sqrt{2x-1}\right)^2 = \left(\sqrt{x-1}+1\right)^2$ **5c.** $\left(\sqrt{2x-1}\right)^2 = x-1+2\sqrt{x-1}+1$ **5d.** $2x-1=x+2\sqrt{x-1}$
5e. $x-1=2\sqrt{x-1}$ **5f.** $x^2-2x+1=4(x-1)$ **5g.** $x^2-6x+5=0$ **5h.** $x=5$ or $x=1$ **5i.** $\{1, 5\}$
Do the Math **1.** Yes **2.** No **3.** No **4.** Yes **5.** $\{-1\}$ **6.** $\{\ \}$ or $\varnothing$ **7.** $\{3\}$ **8.** $\{4, 6\}$ **9.** $\{3, 4\}$ **10.** $\{8\}$

11. $\{\ \}$ or $\varnothing$ **12.** $\{-2, -1\}$ **13.** $\{1\}$ **14.** $\{\ \}$ or $\varnothing$ **15.** $\{3\}$ **16.** $\left\{\dfrac{1}{4}\right\}$ **17.** 240 ft **18.** 27

Section 8.7

Five-Minute Warm-Up 1a. 81 **1b.** 16 **1c.** -256 **2.** $10x^9$ **3.** $\dfrac{3x^2}{4y^3}$ **4.** $-125x^{12}$ **5.** $\dfrac{1}{n^3}$ **6.** 5

Guided Practice 1. index; radicand; root **2.** positive; any real number **3.** principal root **4a.** -3
4b. not a real number **4c.** $\dfrac{2}{3}$ **4d.** 2 **5a.** $2\sqrt[3]{2}$ **5b.** $3n\sqrt[3]{3n^2}$ **5c.** $-3p^2\sqrt[3]{2}$ **6a.** $-2\sqrt[3]{9}$ **6b.** $2z\sqrt[4]{45}$

7a. $\sqrt{144}=12$ **7b.** $\sqrt{-64}$; not a real number **7c.** $-\sqrt{36}=-6$ **8.** $\sqrt[n]{a^m}; \left(\sqrt[n]{a}\right)^m$ **9a.** 216 **9b.** -243 **9c.** 16

9d. not a real number **10a.** $x^{\frac{5}{4}}$ **10.** $\left(2x^2\right)^{\frac{4}{3}}$ **11a.** $3\sqrt[3]{x^2}$ **11b.** $\sqrt[3]{(3x)^2}$ **12a.** $\dfrac{1}{8^{\frac{2}{3}}}=\dfrac{1}{4}$ **12b.** $-\dfrac{1}{16^{\frac{3}{4}}}=-\dfrac{1}{8}$

13a. 36 **13b.** $7x^{\frac{1}{3}}$ **13c.** $-2a^{\frac{1}{2}}$
Do the Math **1.** $\sqrt[3]{27}$ or 3 **2.** not a real number **3.** 3 **4.** 2 **5.** z **6.** $2\sqrt[3]{3}$ **7.** 2 **8.** $5\sqrt[3]{2}$ **9.** $4w^2$ **10.** $2a^3\sqrt[4]{3}$

11. $5\sqrt[3]{p}$ **12.** $\left(\sqrt[4]{-16}\right)^5$, not a real number **13.** $-\dfrac{1}{\left(\sqrt{16}\right)^3} = -\dfrac{1}{64}$ **14.** $\sqrt[4]{(4x)^3} = \left(\sqrt[4]{4x}\right)^3$ **15.** $-4z^{\frac{2}{3}}$ **16.** $(7y)^{\frac{3}{4}}$

17. 36 **18.** $8y^{\frac{1}{4}}$ or $8\sqrt[4]{y}$ **19.** $x^{\frac{4}{5}}$ or $\sqrt[5]{x^4}$ **20.** $9x^2$

Chapter 9 Answers
Section 9.1

Five-Minute Warm-Up **1a.** 13 **1b.** −8 **2a.** $2\sqrt{5}$ **2b.** $5\sqrt{3}$ **3.** $(2z-3)^2$ **4a.** $4\sqrt{2}$ **4b.** ≈ 5.66

5. $3-2\sqrt{2}$ **6.** $4\sqrt{3}$

Guided Practice **1.** $x=\sqrt{p}$ or $x=-\sqrt{p}$ **2.** $\left\{1+2\sqrt{3}, 1-2\sqrt{3}\right\}$ **3.** False **4a.** $n^2=45$ **4b.** $n=\pm\sqrt{45}$

4c. $n=\pm3\sqrt{5}$ **4d.** $\left\{3\sqrt{5},-3\sqrt{5}\right\}$ **5.** $\left\{\dfrac{2\sqrt{15}}{5}, -\dfrac{2\sqrt{15}}{5}\right\}$ **6a.** $(8x-4)^2=20$ **6b.** $8x-4=\pm\sqrt{20}$

6c. $8x-4=\pm 2\sqrt{5}$ **6d.** $8x=4\pm 2\sqrt{5}$ **6e.** $x=\dfrac{4\pm 2\sqrt{5}}{8}$ **6f.** $x=\dfrac{2(2\pm\sqrt{5})}{8}$ **6g.** $x=\dfrac{2\pm\sqrt{5}}{4}$

6h. $\left\{\dfrac{2+\sqrt{5}}{4}, \dfrac{2-\sqrt{5}}{4}\right\}$ **7.** $\left\{5+6\sqrt{2}, 5-6\sqrt{2}\right\}$ **8.** In a right triangle, the square of the length of the

hypotenuse is equal to the sum of the squares of the lengths of the legs; $z^2=x^2+y^2$ **9a.** $4\sqrt{29}$ in.
9b. ≈ 21.5 in.

Do the Math **1.** $\{-9,9\}$ **2.** $\left\{-4\sqrt{3}, 4\sqrt{3}\right\}$ **3.** $\left\{-3\sqrt{3}, 3\sqrt{3}\right\}$ **4.** $\left\{-\dfrac{1}{4}, \dfrac{1}{4}\right\}$ **5.** $\left\{-\dfrac{4\sqrt{7}}{7}, \dfrac{4\sqrt{7}}{7}\right\}$ **6.** $\{-9,-1\}$

7. $\left\{-4-\sqrt{15}, -4+\sqrt{15}\right\}$ **8.** $\{-9,3\}$ **9.** $\left\{\dfrac{1}{4}, \dfrac{1}{2}\right\}$ **10.** no real solution **11.** $\{-14,-2\}$ **12.** $\left\{-\dfrac{2}{3}, 2\right\}$ **13.** $\{6,8\}$

14. $\{-6,6\}$ **15.** $\{-7\}$ **16.** $\left\{-2\sqrt{3}, 2\sqrt{3}\right\}$ **17.** 15 in.

Section 9.2

Five-Minute Warm-Up **1.** $(x+6)^2$ **2.** $\{-1,4\}$ **3.** $x^2-2x+\dfrac{4}{3}$ **4.** $\{-4,5\}$ **5.** $x-9$

Guided Practice **1.** $\left(\dfrac{b}{2}\right)^2$ or $\left(\dfrac{1}{2}b\right)^2$ **2a.** $49; (p-7)^2$ **2b.** $\dfrac{81}{4}; \left(n+\dfrac{9}{2}\right)^2$ **2c.** $\dfrac{4}{9}; \left(z+\dfrac{2}{3}\right)^2$

3a. $p^2-6p=18$ **3b.** 9 **3c.** $p^2-6p+9=27$ **3d.** $(p-3)^2=27$ **3e.** $\sqrt{(p-3)^2}=\pm\sqrt{27}$

3f. $p-3=\pm 3\sqrt{3}$ **3g.** $p=3\pm 3\sqrt{3}$ **3h.** $\left\{3+3\sqrt{3}, 3-3\sqrt{3}\right\}$ **4.** $\left\{-\dfrac{3}{2}+\dfrac{\sqrt{37}}{2}, -\dfrac{3}{2}-\dfrac{\sqrt{37}}{2}\right\}$

5. divide both sides by 3. **6.** $\varnothing$

Do the Math **1.** $144; (x+12)^2$ **2.** $324; (p-18)^2$ **3.** $\dfrac{9}{49}; \left(y+\dfrac{3}{7}\right)^2$ **4.** $\dfrac{25}{4}; \left(z-\dfrac{5}{2}\right)^2$ **5.** $\dfrac{1}{4}; \left(w-\dfrac{1}{2}\right)^2$

6. $\dfrac{1}{64}; \left(r+\dfrac{1}{8}\right)^2$ **7.** $\{-10,4\}$ **8.** $\{2,3\}$ **9.** no real solution **10.** $\left\{3-3\sqrt{3}, 3+3\sqrt{3}\right\}$ **11.** $\{1,2\}$

12. $\left\{\dfrac{-1-\sqrt{13}}{3}, \dfrac{-1+\sqrt{13}}{3}\right\}$ **13.** $\left\{-\dfrac{3}{2}+\sqrt{2}, -\dfrac{3}{2}-\sqrt{2}\right\}$ **14.** no real solution **15.** $\left\{\dfrac{9-\sqrt{53}}{2}, \dfrac{9+\sqrt{53}}{2}\right\}$

16. $\left\{-\dfrac{3}{2}\right\}$ **17.** $\left\{-3\sqrt{2}, 3\sqrt{2}\right\}$ **18.** $\{-2, 15\}$ **19.** $\left\{-4\sqrt{3}, 4\sqrt{3}\right\}$ **20.** $\left\{-4, -\dfrac{1}{2}\right\}$

Section 9.3

Five-Minute Warm-Up **1.** 4 or 2 **2.** $-2+\sqrt{6}$ or $-2-\sqrt{6}$ **3a.** $p^2+3p-15=0$; $a=1; b=3; c=-15$

3b. $2n^2-n-5=0$; $a=2; b=-1, c=-5$ **4.** $\left\{-\dfrac{1}{4}\right\}$

Sullivan/Struve/Mazzarella, *Elementary Algebra*, 2e

Guided Practice 1. $\dfrac{-b \pm \sqrt{b^2 - 4ac}}{2a}$ **2.** standard form **3a.** $2x^2 - x - 4 = 0$; $a = 2, b = -1, c = -4$

3b. $3w^2 + 6 = 0$; $a = 3, b = 0, c = 6$ **3c.** $3y^2 - 6y = 0$; $a = 3, b = -6, c = 0$ **4a.** 8; -2; -3

4b. $n = \dfrac{-b \pm \sqrt{b^2 - 4ac}}{2a}$ **4c.** $n = \dfrac{-(-2) \pm \sqrt{(-2)^2 - 4(8)(-3)}}{2(8)}$ **4d.** 100 **4e.** $\dfrac{2 \pm 10}{16}$ **4f.** $\dfrac{2 + 10}{16}, \dfrac{2 - 10}{16}$

4g. $\left\{\dfrac{3}{4}, -\dfrac{1}{2}\right\}$ **5.** $\left\{3 + 2\sqrt{3}, 3 - 2\sqrt{3}\right\}$ **6.** $\varnothing$ **7.** $b^2 - 4ac$ **8a.** factor **8b.** factor **8c.** quadratic formula or complete the square **8d.** no real solutions **9a.** 169; factor **9b.** -20; no real solutions **9c.** 0; factor

Do the Math 1. $\left\{2 - \sqrt{10}, 2 + \sqrt{10}\right\}$ **2.** $\left\{-5, \dfrac{1}{2}\right\}$ **3.** $\left\{-\dfrac{1}{2}, \dfrac{2}{5}\right\}$ **4.** $\left\{\dfrac{3 - 3\sqrt{5}}{2}, \dfrac{3 + 3\sqrt{5}}{2}\right\}$ **5.** $\{-7, -2\}$ **6.** $\left\{\dfrac{2}{3}\right\}$

7. $\{4\}$ **8.** no real solution **9.** -47; no real solution **10.** 100; factoring **11.** 0; factoring **12.** 32; quadratic formula or complete the square **13.** 0; factoring **14.** -15; no real solution **15. (a)** $1012.50, **(b)** 50

Section 9.4

Five-Minute Warm-Up 1. $b = 4\sqrt{3}$ **2.** ≈ 3.07; ≈ 0.40 **3.** $\left\{2 + \sqrt{2}, 2 - \sqrt{2}\right\}$

Guided Practice 1a. $w + 4$ **1b.** $A = l \bullet w$ **1c.** $117 = (w + 4)w$ **1d.** $w = -13, w = 9$ **1e.** $w = -13$ **1f.** The rectangle has dimensions $9\text{m} \times 13\text{m}$. **2a.** 3 **2b.** $N = \dfrac{2000}{3^2 + 5}$ **2c.** $142.86 **2d.** $1714 **2e.** During the third year, Joel spent $1714 on advertising. **2f.** After 8.66 years, Joel will spend $25/month **3a.** 0.2 sec and 9.5 sec **3b.** 9.7 sec

Do the Math 1. width = 10 ft; length = 15 ft **2.** length = 12 cm ; width = 10 cm

3. height = 15 in.; base = 8 in. **4.** height = 18 m; base = 11 m **5.** longer leg; $\dfrac{5 + 5\sqrt{7}}{2} \approx 9.1$ km; shorter leg;

$x - 5 \approx 4.11$ km **6.** leg = $\dfrac{2 + 2\sqrt{7}}{3} \approx 2.4$ yd; hypotenuse = $\dfrac{1 + 4\sqrt{7}}{3} \approx 3.9$ yd **7.** $-14, -13$ and $13, 14$

8. 10, 11

9. $2 - 3\sqrt{2}$ or $2 + 3\sqrt{2}$ **10.** $2 - \sqrt{3}$ or $2 + \sqrt{3}$ **11a.** 85 ft **11b.** 157 ft **11c.** 0.61 s and 5.14 s **11d.** 6.01 s
12. $(90 - y)°$; $20°$ or $70°$

Section 9.5

Five-Minute Warm-Up 1. $\left\{25, \sqrt{9}\right\}$ **2.** $\left\{25, \sqrt{9}, \dfrac{0}{15}\right\}$ **3.** $\left\{25, \sqrt{9}, \dfrac{0}{15}, -\dfrac{10}{2}\right\}$

4. $\left\{25, \sqrt{9}, \dfrac{0}{15}, -\dfrac{10}{2}, \dfrac{8}{9}, 0.\overline{3}\right\}$ **5.** $\left\{\sqrt{20}, \pi\right\}$ **6.** $\left\{25, \sqrt{9}, \dfrac{0}{15}, -\dfrac{10}{2}, \dfrac{8}{9}, 0.\overline{3}, \sqrt{20}, \pi\right\}$ **7.** $14x^5 + 21x^3 - 7x^2$

8. $6x^2 + 13x - 5$ **9.** 23

Guided Practice 1. -1; $\sqrt{-1}$ **2.** $a + bi$; real numbers; standard; 6; -2 **3a.** $9i$ **3b.** $-2\sqrt{11}\,i$ **3c.** $3\sqrt{3}\,i$

4a. $6 + 2i$ **4b.** $\dfrac{3}{2} - 4\sqrt{3}\,i$ **4c.** $-3 + \sqrt{5}\,i$ **5.** standard **6a.** $-1 - 11i$ **6b.** $-2 - 5\sqrt{6}\,i$ **7a.** -6 **7b.** $-16 - 30i$

7c. $12 + 9i$ **7d.** $16 + 4i$ **8a.** $3 - i$ **8b.** $\dfrac{20}{(3 + i)} \cdot \dfrac{(3 - i)}{(3 - i)}$ **8c.** $\dfrac{20(3 - i)}{9 + 1}$ **8d.** $2(3 - i)$ **8e.** $6 - 2i$

9. $\dfrac{13}{2} - 2i$ **10.** $\left\{\dfrac{-1 \pm 2i}{5}\right\}$

Do the Math **1.** $15-3\sqrt{5}i$ **2.** $-3-\sqrt{3}i$ **3.** $-1-11i$ **4.** -12 **5.** $5+i$ **6.** $5-18i$ **7.** $-35-10i$

8. $-9+3i$ **9.** $28-96i$ **10.** $16+4i$ **11.** $4+5i;41$ **12.** $2+2\sqrt{3}i;16$ **13.** $3+6i$ **14.** $-5-2i$ **15.** $-\dfrac{4}{5}+i$

16. $\dfrac{3}{2}+\dfrac{1}{2}i$ **17.** $\left\{\dfrac{3}{4}-\dfrac{\sqrt{39}}{4}i,\dfrac{3}{4}+\dfrac{\sqrt{39}}{4}i\right\}$ **18.** $\{-6-3i,-6+3i\}$ **19.** $\{-3\sqrt{5},3\sqrt{5}\}$

20. $\{-3-3\sqrt{2},-3+3\sqrt{2}\}$

Chapter 10 Answers
Section 10.1

Five-Minute Warm-Up **1.–2.** See Graphing Answer Section **3.** $\{-8,6\}$ **4.** $\left\{\dfrac{1\pm\sqrt{6}}{5}\right\}$; ≈ 0.7 or ≈ -0.3

5. $x=-2,\ y=3$

Guided Practice **1a.** $y=(-2)^2-(-2)-4=2;(-2,2)$ **1b.** $y=(-1)^2-(-1)-4=-2;(-1,-2)$

1c. $y=(0)^2-(0)-4=-4;(0,-4)$ **1d.** $y=(1)^2-(1)-4=-4;(1,-4)$

1e. $y=(2)^2-(2)-4=2;(2,-2)$ **1f.** See Graphing Answer Section **2.** setting $y=0$ and solving for x

3. setting $x=0$ and solving for y **4.** $x=6,\ x=-3,\ y=18$ **5.** up **6.** down **7.** $x=-\dfrac{b}{2a}$ **8a.** $-1; 2; 8$

8b. down **8c.** $\dfrac{-2}{-2}=1$ **8d.** $y=9$ **8e.** $(1,9)$ **8f.** $x=1$ **8g.** $y=8$ **8h.** $4+4(1)(8)=36$ **8i.** 2

8j. $x=4;\ x=-2$ **8k.** See Graphing Answer Section

Do the Math **1.** See Graphing Answer Section **2.** x-intercepts: $-\dfrac{2}{3},-1$; y-intercept: -2

3. x-intercepts: $-\sqrt{2}\approx-1.41$, $\sqrt{2}\approx1.41$; y-intercept: -2 **4a.** down **4b.** $(0,3)$ **4c.** $x=0$ **5a.** up

5b. $\left(\dfrac{9}{4},-\dfrac{49}{8}\right)$ **5c.** $x=\dfrac{9}{4}$ **6.** See Graphing Answer Section; opens down; vertex: $(0,25)$; x-intercepts: $-5,5$;

y-intercept: 25; axis of symmetry: $x=0$ **7.** See Graphing Answer Section; opens up; vertex: $\left(-\dfrac{3}{2},0\right)$;

x-intercept: $-\dfrac{3}{2}$, y-intercept: 9; axis of symmetry: $x=-\dfrac{3}{2}$ **8.** minimum; 3 **9.** maximum; 7 **10a.** 37.5 ft

10b. 62.25 ft

Section 10.2

Five-Minute Warm-Up **1.–4.** See Graphing Answer Section
Guided Practice **1.** A relation is a mapping that pairs elements of one set with elements of a second set.
2. Domain: {Arkansas, Delaware, Florida, Illinois, New Mexico, New York, Pennsylvania, Texas, Vermont};
Range: {3, 14, 33, 58, 67, 75, 102, 254} **3.** inputs; x **4.** outputs; y **5a.** $\{x\,|\,x$ is any real number$\}$

5b. $\{y\,|\,y\geq-3\}$ **6a.** $\{x\,|-3\leq x\leq 3\}$ **6b.** $\{y\,|-4\leq y\leq 4\}$ **7.** See Graphing Answer Section **7a.** 1; $(-2,1)$

7b. $-2;(-1,-2)$ **7c.** $-3;(0,-3)$ **7d.** $-2;(1,-2)$ **7e.** 1; $(2,1)$ **7f.** $\{x\,|\,x$ is any real number$\}$

7g. $\{y\,|\,y\geq-3\}$

Sullivan/Struve/Mazzarella, *Elementary Algebra*, 2e

Do the Math **1.** {(Sandy, 426–555), (Kathleen, 278–896), (Shawn, 377–204), (Roberto, 368–110)}; Domain: {Sandy, Kathleen, Shawn, Roberto}; Range: {426–555, 278–896, 377–204, 368–110} **2.** {(Christian, 8), (Maria, 11), (Irma, 5), (Dwayne, 11), (Eleazar, 20)}; Domain: {Christian, Maria, Irma, Dwayne, Eleazar}; Range: {5, 8, 11, 20} **3.** See Graphing Answer Section; Domain: {−2, −4, 1}; Range: {3, 2, 1} **4.** See Graphing Answer Section; Domain: {3, −1, 1}; Range: {0, −6, 6} **5.** Domain: {−2, 1, 3}; Range: {−2, 0, 3, 4} **6.** Domain: $\{x \mid x$ is any real number$\}$; Range: $\{y \mid y$ is any real number$\}$ **7.** Domain: $\{x \mid x$ is any real number$\}$; Range: $\{y \mid y \geq 0\}$

8. Domain: $\{x \mid -3 \leq x \leq 1\}$; Range: $\{y \mid -1 \leq y \leq 3\}$

9. See Graphing Answer Section; Domain: $\{x \mid x$ is any real number$\}$; Range: $\{y \mid y$ is any real number$\}$
10. See Graphing Answer Section; Domain: $\{x \mid x$ is any real number$\}$; Range: $\{y \mid y$ is any real number$\}$
11. See Graphing Answer Section; Domain: $\{x \mid x$ is any real number$\}$; Range: $\{y \mid y \leq 0\}$
12. See Graphing Answer Section; Domain: $\{x \mid x$ is any real number$\}$; Range: $\{y \mid y \geq -3\}$

13. See Graphing Answer Section; Domain: $\{x \mid x$ is any real number$\}$; Range: $\{y \mid y \leq \dfrac{33}{4}\}$

14. See Graphing Answer Section; Domain: $\{x \mid x$ is any real number$\}$; Range: $\{y \mid y \geq -\dfrac{21}{4}\}$

Section 10.3

Five-Minute Warm-Up **1a.** 53 **1b.** 5 **2a.** −4 **2b.** $\dfrac{11}{3}$ **3.** $\dfrac{34}{3}$ **4.** $2\sqrt{2}$ **5a.** $\{-2, -1, 2, 3\}$ **5b.** $\{4, 2, 6\}$

Guided Practice **1.** A function is a relation in which each element of the domain (inputs) corresponds to exactly one element in the range (outputs). **2a.** yes; domain: $\{-1, 4, 5\}$; range: $\{-1, 4, 5\}$ **2b.** no
3a. $2x; -2x$ **3b.** no **4a.** yes **4b.** no **4c.** no **4d.** yes **5.** A graph represents a function if and only if every possible vertical line intersects the graph in at most one point. **6.** (c) and (d) only **7a.** −44 **7b.** 0 **8a.** 14 **8b.** −1 **9a.** 4 **9b.** 4 **10a.** 8 **10b.** 29
Do the Math **1.** function; Domain: {English, Math, Psychology, History}; Range: {125, 85, 120}
2. not a function **3.** function; Domain: {3, 2, 0, −1}; Range: {1} **4.** not a function **5.** not a function
6. function **7.** function **8.** function **9.** not a function **10.** function **11a.** 4 **11b.** 10 **11c.** 0 **12a.** 1 **12b.** −2
12c. 3 **13a.** −1 **13b.** −1 **13c.** −1 **14a.** 4 **14b.** 13 **14c.** 8 **15a.** −3 **15b.** −18 **15c.** −13

Graphing Answer Section

Section 1.3 Guided Practice

4. (number line with points at $-\frac{5}{2}$, -0.5, 1.75, 3)

Section 1.3 *Do the Math*

7. (number line with points at $-\frac{5}{4}$, -0.5, $\frac{0}{2}$, $\frac{3}{4}$, 1.5)

Section 2.7 Guided Practice

7b.

	Rate, mph	Time, hours =	Distance, miles
Slower boat	r	2.5	$2.5r$
Faster boat	$r + 4$	2.5	$2.5(r + 4)$
Total			50 miles

Section 2.7 *Do the Math*

5.

	Rate, mph	Time, hours =	Distance, miles
Beginning of trip	r	2	$2r$
Rest of the trip	$r - 10$	3	$3(r - 10)$
Total		5	580

Section 2.8 Warm-Up

7. (number line with points at -3.5, $-\frac{2}{3}$, $\frac{3}{2}$, 3)

Section 2.8 Guided Practice

2. (number line)   **7f.** (number line)

Section 2.8 *Do the Math*

1. $(5, \infty)$ (number line)

2. $(-\infty, 6]$ (number line)

3. $[-2, \infty)$ (number line)

4. $(-\infty, -3)$ (number line)

9. $\{x \mid x < 3\}; (-\infty, 3)$ (number line)

10. $\{x \mid x > 3\}; (3, \infty)$ (number line)

11. $\{x \mid x \le -4\}; (-\infty, -4]$ (number line)

12. $\{x \mid x > -2\}; (-2, \infty)$ (number line)

13. $\{x \mid x \ge 5\}; [5, \infty)$ (number line)

14. $\{x \mid x \ge -1\}; [-1, \infty)$ (number line)

15. $\left\{x \mid x > \dfrac{11}{2}\right\}; \left(\dfrac{11}{2}, \infty\right)$ (number line)

16. $\varnothing$ or $\{\ \}$ (number line)

17. $\{p \mid p \text{ is any real number}\}; (-\infty, \infty)$ (number line)

Section 3.1 Warm-Up

1. (number line with points at -2.5, -1, $\frac{3}{2}$, 3)

Section 3.1 Guided Practice

1.

QII	QI
QIII	QIV

Section 3.1 *Do the Math*

1.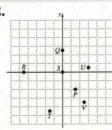

2.

8.

x	y	(x, y)
-3	-17	$(-3, -17)$
1	-1	$(1, -1)$
2	3	$(2, 3)$

9.

x	y	(x, y)
-2	2	$(-2, 2)$
2	-1	$(2, -1)$
4	$-\dfrac{5}{2}$	$\left(4, -\dfrac{5}{2}\right)$

Section 3.2 Warm-Up

6.

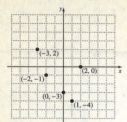

Section 3.2 Guided Practice

1.

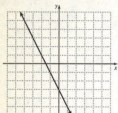

5c.

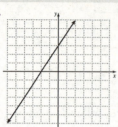

6.

7.

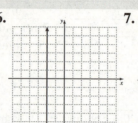

Section 3.2 *Do the Math*

3.

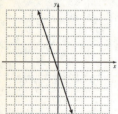

4.

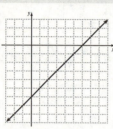

11.

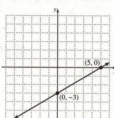

12.

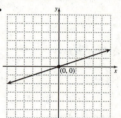

13.

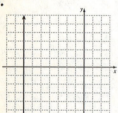

14.

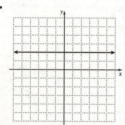

Section 3.3 Guided Practice

9.

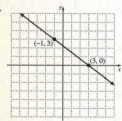

Section 3.3 *Do the Math*

3a., 3b.

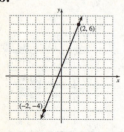

4a., 4b.

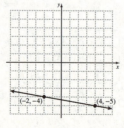

7.

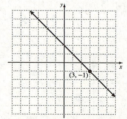

8.

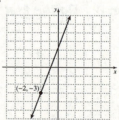

Sullivan/Struve/Mazzarella, *Elementary & Intermediate Algebra*, 2e

Section 3.4 Guided Practice

4c.

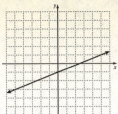

5d.

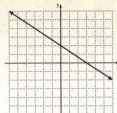

Section 3.4 *Do the Math*

5.

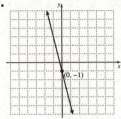

6.

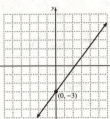

7.

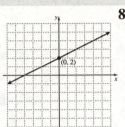

8.

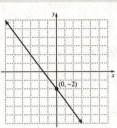

15e.

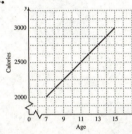

Section 3.5 *Do the Math*

1.

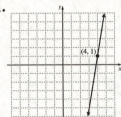

2.

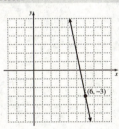

3.

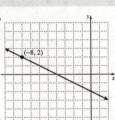

4.

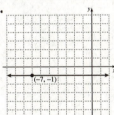

11b.

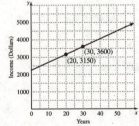

Section 3.6 *Do the Math*

1-3.

Slope of the Given Line	Slope of a Line Parallel to the Given Line	Slope of a Line Perpendicular to the Given Line
$m = 4$	4	$-\dfrac{1}{4}$
$m = -\dfrac{1}{8}$	$-\dfrac{1}{8}$	8
$m = \dfrac{5}{2}$	$\dfrac{5}{2}$	$-\dfrac{2}{5}$

Section 3.7 Warm-Up

4.
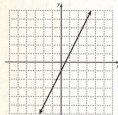

Section 3.7 Guided Practice

6h.
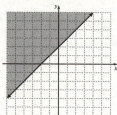

7c.
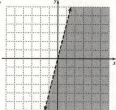

Section 3.7 *Do the Math*

5.
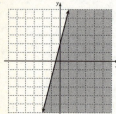

6.
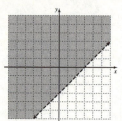

7.
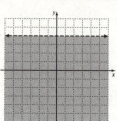

8.
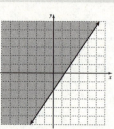

9.
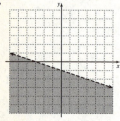

10.
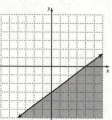

11.
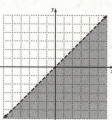

12.
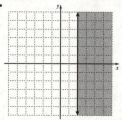

Section 4.1 Warm-Up

1a.
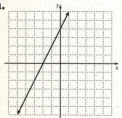

1b.
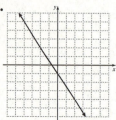

Section 4.1 Guided Practice

2g.
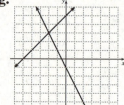

3.
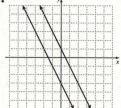

Section 4.1 *Do the Math*

3.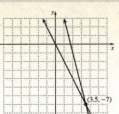

4.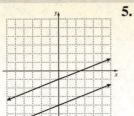

5.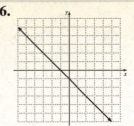

6.

11.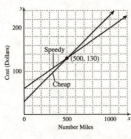

Section 4.4 Guided Practice

4c., 4d.

	Distance (miles)	Rate (mph)	Time (hours)
With the wind	2400	$a + w$	6
Against the wind	2400	$a - w$	8

Section 4.5 Guided Practice

1a–c.

	Number	Cost per Person in Dollars	Amount
Adults	a	7.50	$7.5a$
Students	s	4.00	$4s$
Total	40		202

2a–c.

	Number of coins	Value per Coin in Dollars	Total Value
Quarters	q	0.25	$0.25q$
Dimes	d	0.10	$0.1d$
Total	42		6.75

3a–c.

	Price \$/Pound	Number of Pounds	Revenue
Almonds	6.50	a	$6.5a$
Peanuts	4.00	p	$4p$
Mix	6.00	50	300

4a–c.

	Number of ml	Concentration	Amount of Pure HCl
25% HCl solution	x	0.25	$0.25x$
40% HCl solution	y	0.40	$0.4y$
30% HCl solution	90	0.30	27

Section 4.5 *Do the Math*

1.

	Number	·	Cost per Item	=	Total Value
Bracelets	b		10		$10b$
Necklaces	n		15		$15n$
Total	69				895

2.

	Principal	·	Rate	=	Interest
Savings Account	s		0.0275		$0.0275s$
Certificate of Deposit	c		0.02		$0.02c$
Total	1700				37.75

3.

	Number of Pounds	Price per Pound	=	Total Value
Peanuts	p	5		$5p$
Trail Mix	t	2		$2t$
Total	40	3		$3(40)$

Section 4.6 Warm-Up

3a. 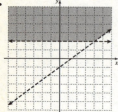 **3b.**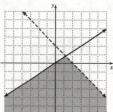

Section 4.6 Guided Practice

7. 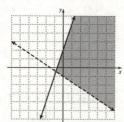 **8d.**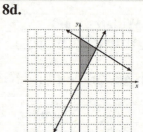

Section 4.6 *Do the Math*

3.   **4.** **5.** 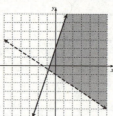 **6.**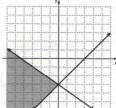

7a.

Number lbs. hamburger (vertical axis), Number lbs. bratwurst (horizontal axis)

Section 6.2 Guided Practice

3a.

Factors whose product is –56	1, –56	2, –28	4, –14	7, –8	8, –7	14, –4	28, –2	56, –1
Sum of factors	–55	–26	–10	–1	1	10	26	55

4a.

Factors whose product is –24	1, –24	2, –12	3, –8	4, –6	6, –4	8, –3	12, –2	24, –1
Sum of factors	–23	–10	–5	–2	2	5	10	23

Section 10.1 Warm-Up

1.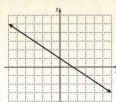

2.

Section 10.1 Guided Practice

1f.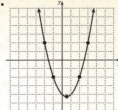

8k.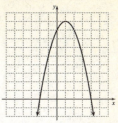

Section 10.1 *Do the Math*

1.

x	y
−3	7
−2	2
−1	−1
0	−2
1	−1
2	2
3	7

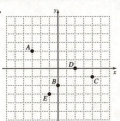

6.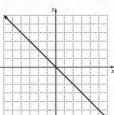

7.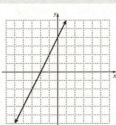

Section 10.2 Warm-Up

1.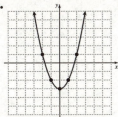

2.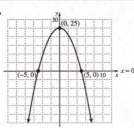

3.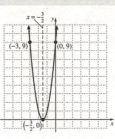

4.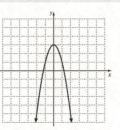

Section 10.2 Guided Practice

7.

Section 10.2 *Do the Math*

3. **4.**

9.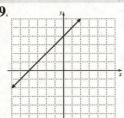

10.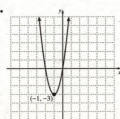

11.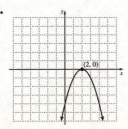

12.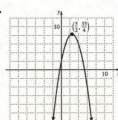

13.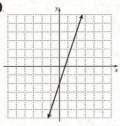

14.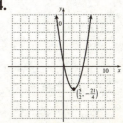